AF412237

Is Economic Growth Sustainable?

This is IEA conference volume no. 148

Is Economic Growth Sustainable?

Edited by

Geoffrey Heal

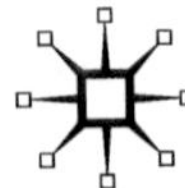

First published 2010 by
PALGRAVE MACMILLAN

Palgrave Macmillan in the UK is an imprint of Macmillan Publishers Limited, registered in England, company number 785998, of Houndmills, Basingstoke, Hampshire RG21 6XS.

Palgrave Macmillan in the US is a division of St Martin's Press LLC, 175 Fifth Avenue, New York, NY 10010.

Palgrave Macmillan is the global academic imprint of the above companies and has companies and representatives throughout the world.

Palgrave® and Macmillan® are registered trademarks in the United States, the United Kingdom, Europe and other countries.

ISBN-13: 978–0–230–23247–1 hardback

This book is printed on paper suitable for recycling and made from fully managed and sustained forest sources. Logging, pulping and manufacturing processes are expected to conform to the environmental regulations of the country of origin.

A catalogue record for this book is available from the British Library.

Library of Congress Cataloging-in-Publication Data
Is economic growth sustainable? / edited by Geoffrey Heal.
    p.   cm. — (IEA conference volume; no. 148)
Includes bibliographical references and index.
Summary: "How can we determine if current growth patterns are sustainable, and what changes do we need to make to make them more so? This volume addresses these issues in a rigorous yet accessible fashion, presenting the current research of some of the world's leading scholars"—Provided by publisher.
ISBN 978–0–230–23247–1 (hardback)
1. Sustainable development.   I. Heal, G. M.
II. International Economic Association.
HC79.E5I8 2010
338.9′27—dc22   2009045180

10 9 8 7 6 5 4 3 2 1
19 18 17 16 15 14 13 12 11 10

Printed and bound in Great Britain by
CPI Antony Rowe, Chippenham and Eastbourne

Contents

List of Tables and Figures

Tables

Figures

List of Contributors

Kenneth J. Arrow was born in 1921, graduated from the College of the City of New York (1940), and received an MA (1941) and PhD (1951) from Columbia University. He has taught at Stanford University, Harvard University, and the University of Chicago, and is currently Professor Emeritus of Economics and of Management Science and Engineering at Stanford University. His principal research fields have been social choice, general equilibrium, economics of uncertainty and information, inventory theory, optimal growth with special reference to environmental constraints, health economics, and the economics of innovation.

James Boyd received his PhD in Applied Microeconomics from the Wharton School, and has served on National Academy of Science and other advisory panels, including most recently the US EPA's Committee on Valuing Ecological Systems and Services. He has been a visiting professor at Stanford University (2007–2008) and Washington University in St. Louis (1996), and was Director of the Energy and Natural Resources Division at Resources for the Future (2002–2007).

Brian R. Copeland is Professor and Head of the Department of Economics at the University of British Columbia. His research has focused on developing analytical techniques to study the interaction between international trade and the environment. He and his colleague Scott Taylor are the authors of the book, *Trade and the Environment: Theory and Evidence*. He was previously co-editor of the *Journal of Environmental Economics and Management* and is currently an associate editor of the *Journal of International Economics*.

Maureen Cropper is Professor of Economics at the University of Maryland, a Senior Fellow at Resources for the Future, and a former Lead Economist at the World Bank. She has served as chair of the EPA Science Advisory Board Environmental Economics Advisory Committee and as past president of the Association of Environmental and Resource Economists. She is a member of the National Academy of Sciences and a Research Associate of the National Bureau of Economic Research. Her research has focused on valuing environmental amenities (especially environmental health effects), on urban land use and transportation and on the evaluation of environmental policies.

Partha Dasgupta was born in Dhaka and educated in Varanasi, India, and Cambridge. He is the Frank Ramsey Professor of Economics at the University of Cambridge, a Fellow of St John's College, Cambridge and Professor of Environmental and Development Economics at the University of Manchester (2008–). He taught at the London School of Economics during 1971–1984 and moved to Cambridge in 1985 as Professor of Economics, where he served as Chairman of the Faculty of Economics in 1997–2001. During 1898–1992 he was also Professor of Economics, Professor of Philosophy and Director of the Program in Ethics in Society at Stanford University, and during 1991–1997 he was Chairman of the Scientific Advisory Board of the Beijer International Institute of Ecological Economics, Stockholm.

Lawrence H. Goulder is the Shuzo Nishihara Professor in Environmental and Resource Economics at Stanford University. He is also a Research Associate at the National Bureau of Economic Research and a University Fellow of the nonprofit research firm Resources for the Future. Much of Goulder's research focuses on economic and environmental impacts of policies to address climate change and pollution from power plants and automobiles. He is also involved in developing more comprehensive accounting methods for improved assessments of the long-run development prospects for various countries. At Stanford Goulder teaches undergraduate and graduate courses in environmental economics and policy, and co-organizes a weekly seminar in public and environmental economics.

Kirk Hamilton is Lead Environmental Economist and Team Leader, Policy and Economics, at the Environment Department of The World Bank, where his current projects include analytical work on the links between poverty and environment, 'greening' the national accounts, and the economics of climate change. He is a co-author of the World Development Report 2010 on development and climate change. Previously senior research fellow at the UK Centre for Social and Economic Research on the Global Environment, Dr. Hamilton has researched and published extensively on growth theory and the economics of sustainable development. He also served as Assistant Director of National Accounts for the government of Canada, where his responsibilities included developing an environmental national accounting program. His degrees include a PhD in Economics and an MSc in Resource and Environmental Economics from University College London, as well as a BSc (Eng.) from Queen's University at Kingston.

Geoffrey Heal, Paul Garrett Professor of Public Policy and Corporate Responsibility at Columbia Business School, is noted for contributions to economic theory and resource and environmental economics. Author of 18 books and about 200 articles, he is a Fellow of the Econometric Society, Past President of the Association of Environmental and Resource Economists, recipient of its prize for publications of enduring quality and a Life Fellow, a member of the Scientific Advisory Board of the Environmental Protection Agency, and a Director of the Union of Concerned Scientists. Recent books include *Nature and the Marketplace*, *Valuing the Future*, and *When Principles Pay*. He chaired a committee of the National Academy of Sciences on valuing ecosystem services, was a Commissioner of the Pew Oceans Commission, and is a Director of Petromin Holdings PNG Ltd. and chairs the Advisory Board of the Coalition for Rainforest Nations. He has been a principal in two start-up companies, one a consulting firm and the other in software and telecommunications.

Kevin Mumford is an assistant professor of Economics at Purdue University. His research examines issues of public policy, such as how fertility responds to child subsidies, the degree of psychological stigma associated with transfer payments, how labor supply responds to government interventions, and the optimal income tax treatment of families with children. At Purdue, he teaches courses in labor economics, public finance, and econometrics. He received a BA in economics from Brigham Young University and a PhD in economics from Stanford University.

Kirsten Oleson is a Post-Doctoral Fellow with Stanford's Masters in Public Policy Program. She teaches the capstone policy analysis practicum, Collective Action, Justice, and a PhD seminar on interdisciplinary research design. She convenes an environmental norms workshop and an environmental ethics working group. Her research is at the interface of environmental ethics, economics, and policy. Kirsten earned her PhD from Stanford's Interdisciplinary Program in Environment and Resources (2007), a Master's in Applied Environmental Economics (Imperial College of London 2005), and degrees in Civil and Environmental Engineering (MSc Technical University of Delft 1998; BSc University of Virginia 1996). Between 1998 and 2003, Kirsten worked at the World Bank supervising environmental assessments of large infrastructure projects.

Elinor Ostrom is Arthur F. Bentley Professor of Political Science and Senior Research Director of the Workshop in Political Theory and Policy Analysis, Indiana University, Bloomington; and Founding Director, Center for the Study of Institutional Diversity, Arizona State University. She is a member of the American Academy of Arts and Sciences, the National Academy of Sciences, and the American Philosophical Society, and a recipient of the Sveriges Riksbank Prize in Economic Sciences in Memory of Alfred Nobel 2009, Reimar Lüst Award for International Scholarly and Cultural Exchange, the Elazar Distinguished Federalism Scholar Award, the Frank E. Seidman Distinguished Award in Political Economy, the Johan Skytte Prize in Political Science, the Atlas Economic Research Foundation's Lifetime Achievement Award, and the John J. Carty Award for the Advancement of Science. Her books include *Governing the Commons* (1990); *Rules, Games, and Common-Pool Resources* (1994, with Roy Gardner and James Walker); *Local Commons and Global Interdependence: Heterogeneity and Cooperation in Two Domains* (1995, with Robert Keohane); *Trust and Reciprocity: Interdisciplinary Lessons from Experimental Research* (2003, with James Walker); *The Commons in the New Millennium: Challenges and Adaptations* (2003, with Nives Dolšak); *The Samaritan's Dilemma: The Political Economy of Development Aid* (2005, with Clark Gibson, Krister Andersson, and Sujai Shivakumar); *Understanding Institutional Diversity* (2005); *Understanding Knowledge as a Commons: From Theory to Practice* (2007, with Charlotte Hess); and *Working Together: Collective Action, the Commons, and Multiple Methods in Practice* (forthcoming 2010, with Amy Poteete and Marco Janssen).

Michael J. Roberts is an assistant professor in the Department of Agricultural and Resource Economics at North Carolina State University. His recent research has focused on the potential effects of climate change and impacts of land conservation and agricultural policies. He has previously held positions as an economist at the US Department of Agriculture and as an adjunct professor in the Paul H. Nitze School of Advanced International Studies at Johns Hopkins University. He teaches economic principles to undergraduates and economic theory to PhD students.

Wolfram Schlenker is an assistant professor in the Department of Economics and the School of International and Public Affairs at Columbia

University and a Faculty Research Fellow at the National Bureau of Economic Research. He received his PhD in Agricultural Economics from the University of California at Berkeley and a Master of Environmental Management from Duke University. His research focuses on the potential effects of climate change on agriculture, sustainable fisheries, as well as optimal decisions under uncertainty.

The Sustainability of Economic Growth

Geoffrey Heal

Is growth sustainable? There is probably not a more important or timely economic question. As the chapters in this volume were completed, an era of economic growth was coming to an end. It ended not because of reasons related to the normal sustainability issues of environmental degradation or climate change, but because of poor assessment of risks and massive indebtedness in the United States. Nevertheless, its ending showed that even without concerns relating to the environment, growth may not be sustainable. But the focus of this book is on longer-term issues, on the sustainability of growing living standards over the long term, decades or even centuries. The focus is on problems that could stop our economies even if we manage our macroeconomic policies perfectly and avoid the errors that have traditionally brought periods of expansion to a halt.

What are these problems that could harm our growth prospects? They are related to the destruction of the natural infrastructure on which our economies depend, of the natural capital that provides a flow of ecosystem services that are essential to our long-run well-being. They include the extinction of species, radical changes to Earth's climate, depletion of natural resources including soils, and many others. These are often the unintended by-products of economic activity, external costs inflicted by economic progress as conventionally measured, now reaching unprecedented levels of destructiveness.

How serious are these external effects? Do we really need to change the ways in which our economies operate to reduce them? And if so, what are the institutional changes that are needed? The three groups of chapters that follow address these three issues. There are two impact studies, providing partial answers to the question "How serious are the problems of environmental degradation?" Then there is a set of papers that define and measure and make operational the idea of sustainability. Finally, there are chapters that study the institutional framework within which we operate and ask how this affects sustainability and how it should be revised.

Maureen Cropper, Kristin Aunan, Pan Xiaochuan, and Zhang Yanshen present one of the first detailed studies of the impact of air pollution in China. Everyone knows that pollution in Chinese cities is bad, and if you've been there you have personal experience of the chemical smogs that envelop urban Chinese populations. But how much do these matter? Cropper et al. answer this in terms of health impacts, and their answer is "a lot." The authors use various methods for estimating and monetizing the health impacts of urban air pollution, summarized by PM_{10} data, and find that these could be as large as 3.8 percent of Chinese gross domestic product.

Michael Roberts and Wolfram Schlenker look at the impact of climate change. While most scientists assume that climate change has the potential to be seriously damaging to human well-being, we have few quantitative studies that meet the highest statistical standards. This is one of them, focusing on climate change and agriculture in the United States. The answer again is that climate change poses a serious threat to agricultural productivity, even in a temperate country like the United States, which is generally thought to be one of the countries least likely to be harmed by climate change. Roberts and Schlenker find that the impact of climate change on US crop yields will be dramatic, reducing them by in the range of 40–50 percent. And this is on the assumption that there will be no reduction in precipitation, which is definitely an optimistic interpretation of climate change scenarios. Driving these results is a highly nonlinear response of plant behavior to ambient temperature: positive up to a certain temperature and then abruptly and sharply negative. Ambient temperatures will cross this threshold in the main US agricultural regions during the 21st century. Roberts and Schlenker point out that the consequences of a 40–50 percent drop in US agricultural yields are far-reaching, even though agriculture is only a small fraction of GDP. It is clearly a very important part of GDP as far as human welfare is concerned, a part that we obviously cannot do without. This is reflected in the fact that there is a vast consumer surplus associated with the consumption of food in rich countries, and that consumers' willingness to pay for food greatly exceeds what they actually pay. This is also reflected in the very inelastic demand for food, which indicates that a small drop in food supply will drive prices up very sharply. Roberts and Schlenker also make the point that the United

States produces about 25 percent by calorific value of most of the world's basic food crops, so that a drop in US production will lead to global shortfalls and price increases. There are now a growing number of studies for developing countries that also show a sharp negative response of food production to climate change (Cline 2007; Guiteras, R. 2007).

These chapters suggest that the environmental impacts of economic growth are serious, serious enough to call into question the appropriateness of continuing our current patterns of economic growth. In other words, this growth may not be sustainable. But what exactly does this mean? How do we measure sustainability? The chapters by Kirk Hamilton; by Kenneth Arrow, Partha Dasgupta, Lawrence Goulder, Kevin Mumford, and Kirsten Oleson; and by James Boyd begin to address this question.

Hamilton develops a theoretical framework for defining and measuring sustainability, based on earlier ideas of Pearce and Atkinson (1993). This is the framework based on the "genuine savings" measure, also known as "adjusted net savings." The basic proposition is that an economy cannot be sustainable unless the total value of its capital stock is increasing. Income comes from wealth, and income cannot be maintained unless wealth is constant, and cannot be increased on a long-term basis unless wealth rises. This wealth measure or capital stock must be very broadly defined to include all stocks that can affect human welfare; so in addition to conventional items such as built capital it has to include human capital and natural capital, the stock of environmental assets that can provide a flow of services (see Barbier and Heal for more discussion). Such assets include obvious physical stocks such as oil and gas reserves, and also less readily measurable but no less important variables such as the state of the climate system.

Hamilton emphasizes an important point, which is that deciding whether an economy's growth is sustainable is making a judgment, a forecast, about the future, in general about the quite distant future. A sustainability measure must be forward-looking, a point that has been noted since a paper in 1961 by Samuelson in which he conjectured that the equivalent of national income in a dynamic economy would have to look at the future flow of consumption; and indeed this is the basis for the genuine savings measure (for a more detailed analysis see Heal and Kristrom 2008). Capital stocks represent the capacity to produce in the future, and their prices should in principle—and here is a real measurement problem—reflect the value of their future products. In practice we have market prices of some capital goods, but not of all, and in particular not of most forms of natural capital. There is also a real doubt that the market prices of forms of capital that are traded fully reflect the values of their future contributions to welfare. These problems notwithstanding, the World Bank has done some remarkable work in evaluating genuine savings for all countries in the world, and Hamilton summarizes these results. According to these results, nonsustainability is mainly a problem of very poor countries and resource-exporters.

Arrow et al. present the results of applying to China and the United States the concepts that Hamilton discusses in his chapter. By restricting their attention to just two countries for which reasonable economic statistics are available (though many would question the accuracy of Chinese economic statistics), they are able to conduct a more detailed implementation than the World Bank could in its study comparing a large number of countries, including many developing countries for which only limited environmental data is available. They use a more sophisticated approach to the measurement of human capital and technological progress, and also to the measurement of the depletion of natural capital, which would seem to be one of the main negative by-products of economic growth, particularly in China given the results of the study by Cropper and her colleagues. They also estimate the consequences of the emission of greenhouse gases.

The conclusions reached with respect to the sustainability of growth in the United States and China will surprise many people: both emerge as highly sustainable, with total wealth per capita growing at 1.8 percent annually in the United States and 5.05 percent in China. Perhaps in the case of China this reflects a gross savings rate in excess of 30 percent of GDP, a rate so high that it would take massive environmental degradation to overcome its contribution to wealth formation. Nevertheless, massive environmental degradation is precisely what many environmentalists associate with China. Either they were misjudging the situation or the Arrow et al. calculations are misleading. In the case of the United States, widely regarded by many environmentalists as the paradigm of unsustainable consumption paths, it is not so clear what is generating the positive outcome. Certainly in the last 30 years, following a burst of environmental legislation during the Johnson and Nixon presidencies, the United States has greatly improved the quality of its air and water, and shown greatly increased concern for species conservation. And it is also the world's main source of technological innovation. But it is still a major emitter of greenhouse gases: indeed there is an irony in the fact that Arrow et al. find the world's two largest emitters of greenhouse gases to be eminently sustainable by the comprehensive wealth criterion. They use rather conservative estimates of the impact of climate change on the United States, those of Nordhaus and Boyer, which are quite at variance with the results of Roberts and Schlenker in Chapter 2, and they also use a relatively low price for greenhouse gas emissions, $50 per ton of carbon (equivalent to roughly $14 per ton of CO_2). These choices could explain some of the unexpectedly positive outcomes, but probably no more than a small part. Either the results are basically correct in their implications, or there is a more fundamental issue with the method chosen for measuring sustainability.

What this might be can be seen in Hamilton's remark that assessing sustainability is forecasting the future. Central to the adjusted net savings approach to sustainability is the proposition that a country's welfare is rising over a short interval of time if and only if at the start of that interval the value of total wealth is rising. So rising wealth, which is what Arrow et al. find, is an indicator that at the moment, or over the near future, welfare is rising. But it is not an indicator that over the next several decades this will continue to be true. One can only judge whether this is the case by running simulations of models over that time horizon. So it may be that while the United States and China are currently meeting the criteria for sustainability, a continuation of their present paths will lead them into unsustainable behavior within decades. Maybe we ideally need a more forward-looking criterion for sustainability.

Boyd's chapter is a further contribution to the measurement of sustainability. Natural capital is an important asset and provides as a return a flow of services, ecosystem services. These contribute to human welfare and to measure this contribution accurately we need to measure the ecosystem services. We also need to be able to do this if we are to assess whether the flow of ecosystem services is constant or is falling, important in evaluating sustainability. Boyd points out that it is not obvious what units we should use for ecosystem services, nor even what exactly the services are. Manufactured goods come in neat well-defined units—cars, DVDs, TVs, etc.—but ecosystem services do not. If we think of the services provided by a watershed, water purification, and stream-flow management (see Heal), then it is not immediately obvious how we quantify these. Likewise for the ecosystems that contribute to climate stability. Yet we need to do this if we are to measure them and assess their trends over time. Boyd's chapter is about constructing an index of ecosystem services. He makes the important accounting point that not all welfare-significant ecosystem services should be counted in the total of such services provided by natural capital: some are final goods or services but others are intermediate goods and services that contribute to something else that affects welfare and whose value is therefore captured in the value of that other commodity. For example, some aspects of climate stability contribute to food production and so they are intermediate goods, whose value is captured in the value of food. Other aspects contribute more directly to human welfare and so are final goods that need to be valued as public goods.

The final three chapters, by Elinor Ostrom, Brian Copeland, and Geoffrey Heal, address institutional incentives for sustainable behavior. So far we have discussed the impacts of environmental degradation on human welfare, and the definition and measurement of sustainability. Now we need to see what institutional frameworks will actually lead to more sustainable outcomes.

Ostrom focuses on the management of common property resources, which, of course, include many of the environmental resources that constitute natural capital and whose depletion threatens the sustainability of our economies. Fisheries, forests, water systems, and even the climate system as a whole can usefully be seen as common property resources. What we know about these is that with open access the Nash equilibrium usage levels will be excessive, leading to depletion and degradation. Ostrom contrasts this conventional theoretical wisdom with results of experimental studies and of field studies of the management of common property resources, mainly fisheries, forests, and irrigations systems. As she states, the standard game-theoretic model is "value-free": agents just act to maximize their utilities without reference to any social norms or institutional structures. In practice, she notes, we rarely observe this outcome in either experiments or actual management practices. People are smart enough that they recognize the limitations of narrow self-interest and the gains from cooperation. This occurred in the lab experiments that she reports, with subjects taking advantage of opportunities to meet and talk to devise collaborative approaches and then sanctioning those who violate them. The same is true in real-world management of common property resources: most social groups have recognized the need to establish norms of behavior that prevent overexploitation and also establish ways of sanctioning noncompliance. What most of the field and lab studies make clear is that achieving outcomes that improve on the Nash equilibrium depends on there being a relatively stable population of resource users. In this case they can internalize the norms and are vulnerable to sanctions: the pressures of social groups do not so easily control transient or widely dispersed populations.

Copeland deals with another institution that has a great impact on how we interact with the natural environment—the international trading system. There has been extensive discussion of the effect of globalization on trade, much of it not well based on the theoretical and empirical literature on this issue. Copeland reviews this literature, and makes clear the subtlety of questions about the impact of trade expansion on the environment. A move to freer trade, as he notes, will have many different consequences, each of which may affect the environment in several ways. He distinguishes the composition, technique, and direct effects of trade on the environment. The composition of production is altered in response to trade, the techniques used to produce a given good may also change in the face of international competition, and then trade has a direct environmental impact through the use of fossil fuels to move goods internationally and the possible introduction of exotic species. Changes in the composition of output and techniques used in production can clearly have environmental impacts, positive or negative.

A much-discussed topic in the area of trade and environment is the "pollution haven" hypothesis, the suggestion that producers will relocate to countries or regions with low environmental standards, or that strong environmental standards put a country's producers at a competitive disadvantage. The former assertion is the strong version of the hypothesis and the latter the weak one. Copeland reviews both theoretical and empirical support for these. Neither is overwhelming: the prerequisites for these effects to happen in a theoretical model are strong, and the empirical evidence is unclear, though with an emerging presumption in support of the weak version. This presumption is based largely on studies of location choices within the United States.

An important question in analyzing the impact of trade is whether economic activity merely pollutes, or whether it adversely affects the stock of natural capital, thus degrading either production possibilities or consumer welfare or both. This latter case leads to a richer and more complex set of models, and in these an important issue is whether pollution is generated by production or consumption. This distinction affects whether local producers gain or lose form tighter environmental standards: they do not lose and may even gain from stricter environmental policies if pollution is consumption-based.

Copeland summarizes this complex topic as follows:

> The evidence to date is that while growth and capital accumulation put significant pressure on the natural environment, there is as yet

little convincing evidence that openness to trade and investment *per se* increase environmental damage, on average. However, it is important to keep in mind that behind these averages lie many individual cases where trade will have significant effects on local communities. Trade can threaten the sustainability of renewable resources when the management regime is weak, and the depletion of such resources can have long-lasting negative effects on communities.

The final chapter of the volume is my own, which looks at the incentives that corporations face to behave sustainably. Corporations are, of course, the main actors in this drama. They mine, deforest, pollute, and produce polluting goods. Generally this is not out of ill-will but in response to the perceived demands of their customers. They are imposing external costs on third parties because there is no legal restriction on doing so and this is the least cost way of producing, allowing them to keep prices down. In short, market forces rather than malevolence drive them in their environmental impacts. My point of departure is that in some cases this seems to be changing: in a growing number of cases, corporations are voluntarily internalizing their external costs and incurring costs to avoid environmentally damaging behavior. Several factors could be contributing to this. One is a growing tendency for third parties subject to external costs to sue for redress, invoking the polluter pays principle. Another is the growth of socially responsible investment (SRI): SRI funds avoid shares in corporations with poor environmental policies, potentially (and in some cases actually) affecting their stock market performance. And to a growing extent, consumers are concerned about the environmental provenance of their purchases. People are beginning to bring their values to their investment and purchasing decisions. So being green is to a growing extent being seen as a corporate virtue that can pay for itself.

Overall the chapters in this book indicate that the costs of unsustainable behavior are high, dangerously so, and that we know enough about defining and measuring sustainability, and about the institutional framework that can support it, to make moving to a more sustainable economy a feasible imperative.

References

Barbier, Edward and Geoffrey Heal. 2006. Valuing ecosystem services: A challenge to policy, *Economists' Voice* (July 2006), www.bepress.com/ev.
Cline, William. 2007. *Global Warming and Agriculture: Impact Estimates by Country*. Washington, DC: Peterson Institute for International Economics.

Guiteras, Raymond. 2007. The impact of climate change on Indian agriculture. Working Paper, Department of Economics, MIT.

Heal, Geoffrey, 2000. *Nature and the Marketplace*, Island Press.

Heal, Geoffrey and Bengt Kristrom. 2008. A note on national income in a dynamic economy, *Economics Letters* 98: 2–8.

Pearce, David and Giles Atkinson. 1993. Capital theory and the measurement of sustainable development: An indicator of weak sustainability, *Ecological Economics* 8: 103–8.

Samuelson, Paul. 1961. The evaluation of 'social income': Capital formation and wealth, Chapter 3, 32–57, in Lutz, F.A. and Haigh, D.C. (editors) *The Theory of Capital*, St. Martin's Press, New York.

1
What Are the Health Effects of Air Pollution in China?*

Maureen Cropper

1 Introduction

China's rapid economic growth, accompanied by industrialization and rapid urbanization, has come at a high environmental price: in 2003 over 50 percent of China's urban population was exposed to annual average PM_{10} levels in excess of $100\,\mu g/m^3$—twice the U.S. standard. The problem of particulate air pollution in China is partly the result of large reserves of high-sulfur coal. China has the world's third largest coal reserves, and over 70 percent of the energy consumed in China is from coal. Approximately half of the coal consumed is burned by industry, often in small boilers,[1] which makes the problem of pollution control difficult. It is also the case that meteorological factors predispose cities in northern China to poor air quality (Pandey 2006).

Evaluating policies to control particulate air pollution through the use of cost-benefit analysis requires that analysts be able to quantify the health effects of air pollution. In the United States, regulatory impact analyses are routinely prepared by the U.S. Environmental Protection Agency (USEPA) to compare the benefits of reducing air pollution to the costs. This requires measuring population exposures to air pollution, relating ambient pollutant concentrations to health endpoints and valuing these endpoints. Specifically, regulatory impact analyses predict ambient population exposures with and without the regulation, evaluate health impacts in both cases, and monetize the health gains associated with the regulation.[2]

The purpose of this chapter is to calculate the health damages associated with particulate air pollution in urban areas of China and to monetize these damages. The analysis is conducted for 660 cities in China, and the results are aggregated to the provincial level.[3] Specifically, we compare annual average PM_{10} levels in 2003 with a reference level ($15\,\mu g/m^3$) of pollution to compute the total health damages associated with particulate pollution, in the spirit of Global Burden of Disease calculations (Cohen et al. 2004). It should, however, be emphasized that the methods presented in this chapter can also be used to determine the benefits of smaller reductions in ambient PM_{10} from 2003 levels, such as would be achievable by realistic pollution control measures.

Although our task would seem to be straightforward, finding appropriate concentration-response functions for China is difficult. In the United States, the majority of the health benefits associated with reductions in particulate come from reductions in premature mortality. Reductions in premature mortality typically comprise over 80 percent of the monetized benefits of air pollution regulations and over 80 percent of the quality-adjusted life years (QALY) associated with air pollution regulations.[4] The USEPA relies on a prospective cohort study by Pope et al. (2002) to compute the impacts of long-term exposure to particulate matter (PM) on deaths in the United States. The World Health Organization (WHO) has also relied on the Pope et al.'s study in computing the burden of disease associated with outdoor air pollution (Cohen et al. 2004).

There is, however, a problem in extrapolating the relative risk function in Pope et al. (2002), which was estimated based on U.S. data, to the much higher air pollution levels experienced in Chinese cities. The WHO team dealt with this problem in their base case by assuming that the risk of death because of PM exposure does not increase after a PM_{10} level of $100\,\mu g/m^3$—it is the same at $150\,\mu g/m^3$ as at $100\,\mu g/m^3$. This assumption is implausible, and erroneously implies that there are no health benefits from reducing PM_{10} levels from 150 to $100\,\mu g/m^3$. We handle the problem, as did Cohen et al. (2004) in a sensitivity analysis, by using a log-linear relative risk function reestimated using data from Pope et al. for this chapter.[5] This relative risk function has the fortuitous property that it agrees with the results of Chinese epidemiological studies at $150\,\mu g/m^3$ of PM_{10}. Concentration-response functions from the Chinese epidemiological literature are used to quantify cases of chronic bronchitis and respiratory and cardiovascular hospital admissions associated with PM_{10}.

Valuing health effects presents a problem in China, just as it does in the United States. The preferred economic approach to valuing mortality risks is to estimate what people would pay for reductions in risk of death. These estimates are typically summed over risk changes that, together, add to one statistical life. Few estimates of the Value of a

Statistical Life (VSL) are available for China. The official approach to valuing mortality risks in China is the Adjusted Human Capital (AHC) approach, which values an avoided death by the present discounted value of per capita GDP computed over remaining life expectancy. This represents an important departure from the traditional human capital approach. Because the use of foregone earnings would assign a value of zero to the lives of the retired and the disabled, the AHC approach avoids this problem by assigning the same value—per capita GDP— to a year of life lost by all persons, regardless of age. For this reason, the AHC approach can perhaps be viewed as a social statement of the value of avoiding premature mortality. We use both the AHC approach and estimates of the VSL obtained by Krupnick et al. (2006) in valuing mortality risks.

The chapter is organized as follows. Section 2 provides information on ambient PM_{10} and SO_2 levels in Chinese cities in 2003. Section 3 discusses the concentration-response functions used in our analysis and uses them to calculate cases of premature mortality and morbidity associated with ambient PM_{10} levels in Chinese cities. Section 4 values these health effects. Section 5 concludes.

2   Population exposure to air pollution

Ambient particulate matter, nitrogen oxides and sulfur dioxide are currently monitored in 341 Chinese cities. Table 1.1 classifies the annual average readings in these cities in 2003 and 2004 according to

Table 1.1   Distribution of PM_{10} and SO_2 levels in Chinese Cities, 2003 and 2004

Distribution of PM_{10} levels	% of Cities	
	2003	2004
$PM_{10} \leq 100\,\mu g/m^3$	46	47
$100 < PM_{10} \leq 150\,\mu g/m^3$	33	39
$PM_{10} > 150\,\mu g/m^3$	21	14
Distribution of SO_2 levels		
$SO_2 \leq 60\,\mu g/m^3$	74	74
$60 < SO_2 \leq 100\,\mu g/m^3$	14	17
$SO_2 > 100\,\mu g/m^3$	12	9

Source: *China Environmental Yearbooks* 2004 and 2005.

Chinese air quality standards. According to Chinese air quality standards, acceptable air quality (Class II or better) implies annual average $PM_{10} < 100\,\mu g/m^3$ (twice the U.S. standard), annual average $SO_2 < 60\,\mu g/m^3$ and $NO_x < 50\,\mu g/m^3$ (both more stringent than the U.S. standard). In 2003 46 percent of Chinese cities met either Class I ($PM_{10} < 40\,\mu g/m^3$) or Class II ($40\,\mu g/m^3 < PM_{10} < 100\,\mu g/m^3$) standards for PM_{10}.

Thirty-three percent of cities had annual average PM_{10} levels between 100 and $150\,\mu g/m^3$ (Class III), and 21 percent of cities had worse than Class III PM_{10} levels (annual $PM_{10} > 150\,\mu g/m^3$). In contrast, annual average ambient SO_2 concentrations exceeded the Class-II standard ($60\,\mu g/m^3$) in only 26 percent of the cities. Annual average NO_2 concentrations in all monitored cities (not shown) met the Class-II standard.

Table 1.1 suggests that PM_{10} is indeed the pollutant of greatest concern in China. Appendix Table A.1 shows the percentage of the urban population exposed to different classes of PM_{10} levels in the 30 provinces of mainland China. Figure 1.1 maps the percentage of urban population exposed to Class III and > Class III PM_{10} levels. Over half of the urban population in China is exposed to annual average PM_{10} levels greater than or equal to $100\,\mu g/m^3$. Over 11 percent are exposed to PM_{10} levels in excess of $150\,\mu g/m^3$, which is three times the U.S. annual average standard. The provinces with the largest percentage of people exposed to PM_{10} levels greater than or equal to $100\,\mu g/m^3$ are generally in the north and central parts of the country (Beijing, Tianjin, Ningxia, Shanxi, Hebei). The provinces with the highest numbers of people exposed are Jaingsu, Sichuan, Henan, and Hubei.

3 Quantifying the health effects of PM_{10}

There is a large international literature showing associations between particulate matter and mortality and morbidity. Time-series or episodic studies have been used to measure the impact of short-term exposures on mortality rates, on the incidence of heart attacks and strokes, and on hospital admissions for respiratory and cardiovascular disease. Cross-sectional studies use variation in air pollution levels across cities to measure the effects of long-term exposures to air pollution on mortality and on the incidence of chronic illness, such as chronic bronchitis.

Ideally, we would like to use studies conducted in China to examine the effects of long-term exposure to PM_{10} on mortality and chronic illness (e.g., chronic bronchitis and asthma) and the impacts of short-term exposures on acute illness. Although the literature on air pollution

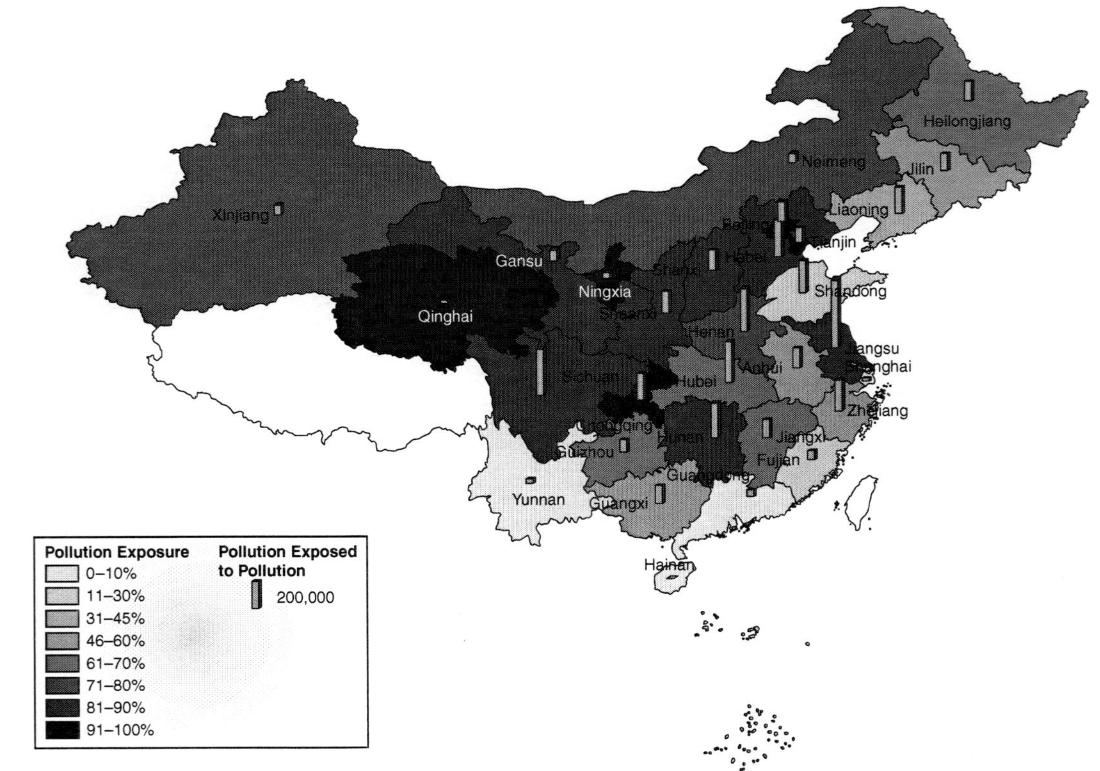

Figure 1.1 Urban population exposed to Class III and > Class III PM$_{10}$ levels, 2003

epidemiology in China is growing, it is necessary to rely on results from the international, as well the Chinese literature. Our choice of concentration-response functions for mortality and morbidity are described below.

3.1   Concentration-response functions for mortality

3.1.1   Time-series studies

There are several dozen studies (Samet et al. 2000; Schwartz et al. 1996) that relate daily variation in air pollution levels within a city to daily deaths, including at least nine such studies in China (Chang et al. 2003; Dong et al. 1995; Gao et al. 1993; Kan and Chen 2003; Kan et al. 2004; Venners et al. 2003; Wang et al. 2003; Xu et al. 1994; Xu et al. 2000). These studies correlate daily variation in the major air pollutants (PM_{10}, $PM_{2.5}$, SO_2, NO_x), measured on day $t - n$, where n is the number of days preceding death, with deaths on day t. Separate equations are often estimated for specific causes of death—respiratory illness and cardiopulmonary disease—and for separate age groups. In addition to controlling variations in air pollution, the studies control daily variation in weather conditions, as well as seasonal factors.

Time-series studies have the advantage of reducing problems associated with confounding variables. Since population characteristics (e.g., smoking habits, occupational exposures, and health habits) are basically unchanged over the study period, the only factors that vary with daily mortality are environmental and meteorological conditions. A disadvantage of time-series studies is that they capture only the impact of short-term peaks in exposure on mortality. Thus, they cannot capture the impact of cumulative exposure to pollution on premature mortality. For this reason, time-series studies have not been used, either by the USEPA or by the WHO (Cohen et al. 2004), to measure the impacts of exposure to air pollution on risk of death.

3.1.2   Cross-sectional studies

To study the impact of long-term air pollution exposures on health, researchers rely primarily on cross-sectional variation in air pollution levels across cities. Two types of long-term exposure studies have found statistically significant associations between mortality rates and particulate matter levels. The first type uses an ecologic cross-sectional study design in which mortality rates for various locations at a single point in time are analyzed to determine if there is a statistical correlation with average air pollutant levels. Such studies have consistently found higher mortality rates in cities (or parts of cities) with higher average levels of particulate matter. Attempts are made to control confounding factors that might be correlated with air pollution levels (such as occupation or migration); however, concerns persist about whether these studies have adequately controlled such factors.

A second type of long-term exposure study involves a prospective cohort design in which a sample is selected and followed over time in each location. Dockery et al. (1993) published results for a 15-year prospective study based on samples of individuals in six U.S. cities. In 1995 Pope et al. (1995) published results of a 7-year prospective study based on samples of individuals in 151 cities in the United States. The Pope et al. study added measurements of air pollution levels (fine particles in 50 cities and sulfates in 151 cities) to data on approximately 500,000 individuals in a prospective cohort assembled by the American Cancer Society. Associations were reported between all-cause mortality and particles and between cardiovascular mortality and particles. The results of this study, based on a longer follow-up period, also show an association between PM and lung cancer (Pope et al. 2002).[6]

In the Chinese literature there are no prospective cohort studies of the effects of air pollution on mortality and there are only two cross-sectional studies that reflect the effects of long-term air pollution exposure on mortality (Jing et al. 1999 and Xu et al. 1996), which were conducted in Benxi and Shenyang. Tables 1.2 and 1.3 summarize salient facts about the Dockery et al. (1993) and Pope et al. (1995, 2002) studies and the two Chinese studies. Because three different size fractions of particles were used in the various studies ($PM_{2.5}$, PM_{10}, and total suspended particulates (TSP)), we convert all measurements to PM_{10} for comparability.[7]

Table 1.2   Summary of cohort studies in the United States

Authors	Year	Locations	Pollutants	Concentration Ranges	Study Design
Dockery et al.	1993	U.S. 6 cities	PM_{10}	$18.2 \sim 46.5\,\mu g/m^3$	Cohort study
Pope et al.	1995	U.S. 61 cities	$PM_{2.5}$	$9.0 \sim 33.5\,\mu g/m^3$	Cohort study
Pope et al.	2002	U.S. 61 cities	$PM_{2.5}$	Mean $= 17.7\,\mu g/m^3$	Cohort study

Source: Dockery et al. (1993); Pope et al. (1995); Pope et al. (2002).

Table 1.3 Summary of ecological studies in China

Authors	Year	Locations	Pollutants	Concentration Ranges	Study Design
Jing et al.	1999	Benxi	TSP	$290 \sim 620\,\mu g/m^3$	Cross-sectional ecological study
Xu et al.	1996	Shenyang	TSP	$353 \sim 560\,\mu g/m^3$	Cross-sectional ecological study

Source: Jing et al. (1999); Xu et al. (1996).

Table 1.4 Summary of results of long-term exposure studies (PM_{10})

Authors	Health End Points	Beta	Std Error
Dockery et al. (1993)	All Cause	0.82	0.28
	Lung Cancer	1.11	0.94
	Cardiopulmonary	1.11	0.37
Pope et al. (1995)	All Cause	0.38	0.09
	Lung Cancer	0.07	0.32
	Cardiopulmonary	0.66	0.14
Pope et al. (2002)	All Cause	0.24	0.10
	Lung Cancer	0.47	0.21
	Cardiopulmonary	0.34	0.13
Jing et al. (1999)	All Cause	0.15	0.06
	Chronic Obstructive Pulmonary Disease (COPD)	0.43	0.17
	Cardiovascular Disease	0.43	0.14
	Cerebrovascular Disease	0.15	0.07
Xu et al. (1996)	All Cause	0.12	0.01
	COPD	0.13	
	Cerebrovascular Disease	0.12	
	Cardiovascular Disease	0.07	

Note: Beta represents the percentage change in the health end point associated with a $1\,\mu g/m^3$ change in PM_{10}.
Source: Dockery et al. (1993); Pope et al. (1995); Pope et al. (2002); Jing et al. (1999); Xu et al. (1996).

Table 1.4 summarizes the results of PM_{10} on premature mortality reported in the five studies. In each study the relative risk (RR) of death associated with a PM_{10} concentration of C_1 compared to a concentration of C_0 is given by

$$RR = \exp(\beta C_1) / \exp(\beta C_0) = \exp(\beta \Delta C). \tag{1}$$

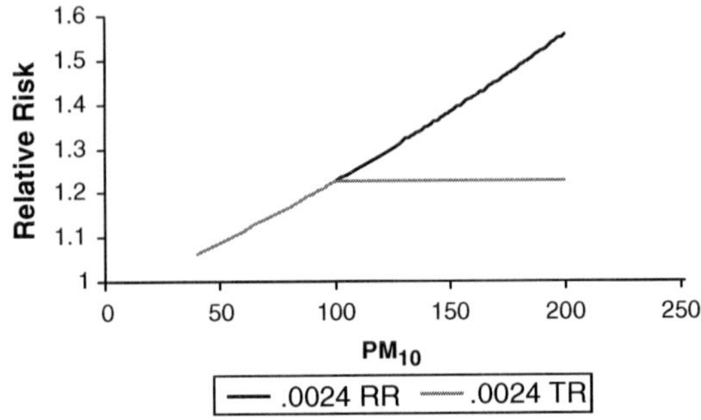

Figure 1.2　Relative risk from Pope et al. (2002) and truncated relative risk
Source: Authors' calculations.

For example, in Pope et al. (2002), $\beta = 0.0024$ for all-cause mortality, implying that the relative risk of dying at PM_{10} level of $50\,\mu g/m^3$ relative to $40\,\mu g/m^3$ PM_{10} is 1.0243.[8] $\beta/100$ (labeled Beta in Table 1.4) is approximately the percentage change in the mortality rate associated with a $10\,\mu g/m^3$ change in PM_{10}. Because (1) is approximately linear in C_1, the risk of dying at a PM_{10} level of C_1 relative to reference level of $15\,\mu g/m^3$ PM_{10} appears as shown in Figure 1.2, where the grey line plots the relative risk of death implied by the Pope et al. (2002) function for all-cause mortality relative to a reference level of $15\,\mu g/m^3$ PM_{10}.

Table 1.4 shows that—holding cause of death constant—the relative risk estimated by U.S. cohort studies is higher than the relative risk estimated by Chinese studies, which were conducted at much higher levels of PM. A meta-analysis of the two Chinese studies implies $\beta = 0.0012$ for all-cause mortality, whereas the corresponding $\beta = 0.0038$ for Pope et al. (1995) and $\beta = 0.0024$ for Pope et al. (2002). Although the relative risk function used in epidemiological studies (equation (1)) is approximately linear in PM concentration, there are indications, from both cross-sectional and time-series mortality studies, that relative risks decline with PM levels. One problem in relying exclusively on the Chinese studies is that the air pollution levels at which effects were measured in these studies are much higher than the air pollution currently experienced in many cities; hence, using these studies at lower air pollution levels may understate the impacts of PM_{10}. On the other hand, reliance on U.S. studies yields implausible results at the air pollution levels currently observed in many Chinese cities.

The implications of relying on U.S. studies are apparent from Figure 1.2. The Pope et al. (2002) relative risk (RR) function reaches 1.38 at a concentration of $150\,\mu g/m^3$, implying (as explained below) that 28 percent of deaths at this concentration (relative to $15\,\mu g/m^3$) are attributable to air pollution. This is clearly an implausible result. WHO (Cohen et al. 2004) dealt with this issue in their base case by assuming that the RR function becomes horizontal at approximately $100\,\mu g/m3$ of PM_{10}, as shown in pink on the graph. This assumption, however, implies that there are no health benefits from reducing PM_{10} from 150 to $100\,\mu g/m^3$.

A compromise solution is to assume that exposure is linear in the logarithm of PM_{10} (Cohen et al. 2004; Ostro 2004), implying that the relative risk function is given by

$$RR = \exp(\alpha + \gamma \ln\ C)/\exp(\alpha + \gamma \ln 15) = (C/15)\gamma. \qquad (2)$$

To estimate the implications of PM for mortality we use a log-linear relative risk function estimated using data from the ACS cohort study.[9] The all-cause mortality data from the ACS study (Pope et al. 1995, 2002) for the 1982–2000 follow-up period were used to estimate a random effects, Cox proportional hazard model with errors clustered by metropolitan area. $PM_{2.5}$, measured in 1979–82, entered the equation in log form. The estimated $\gamma = 0.073$ (S.E. $= 0.028$).

This relative risk function (labeled Ostro RR) is plotted in Figure 1.3. It coincides with the RR function based on (1) with $\beta = 0.0012$ at $150\,\mu g/m^3$ and yields higher relative risks at lower PM_{10} levels. Figure 1.3 compares this RR function with the RR function implied by equation (1) with $\beta = 0.0012$. Equation (2) is used to compute the relative risks of PM_{10} concentrations on all-cause mortality in this chapter.[10]

3.2   Concentration-response functions for morbidity

In estimating the impact of PM_{10} on morbidity, we rely exclusively on Chinese studies and focus on endpoints—chronic bronchitis, respiratory hospital admissions, and cardiovascular hospital admissions—that are examined in the international literature. Hospital admission studies are not easily transferred from one country to another as they depend on country-specific features of the health care system. In the case of chronic bronchitis, there are more Chinese studies than are available to estimate the long-term impacts of PM exposure on mortality.

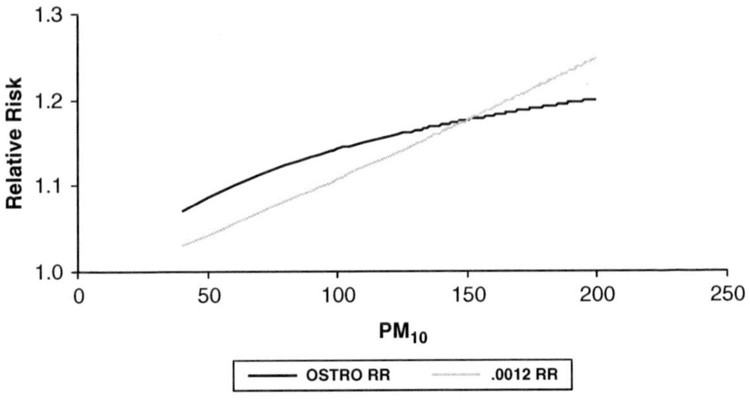

Figure 1.3   Relative risk functions based on U.S. and Chinese studies
Source: Authors' calculations.

3.2.1   *Exposure-response coefficients for hospital admissions*

Few studies have been carried out in China addressing hospitalization associated with air pollution (Aunan and Pan 2004; HEI 2004). We apply the functions derived in Aunan and Pan (2004) to estimate the number of annual excess cases of hospital admissions for cardiovascular diseases and respiratory diseases. The functions are based on two time-series studies in Hong Kong and indicate a 0.7 percent (S.E. $= 0.02$) increase in hospital admissions because of cardiovascular diseases per $10\,\mu g/m^3$ PM_{10} and a 1.2 percent (S.E. $= 0.02$) increase in hospital admissions because of respiratory diseases per $10\,\mu g/m^3$ PM_{10}. The relative risks for hospital admissions are given by (1) with the values of $\beta = 0.0007$ and $\beta = 0.0012$, respectively. The number of workdays lost associated with each hospital admission is calculated as the average length of the hospital stay.

3.2.2   *Exposure-response coefficients for chronic bronchitis*

Aunan and Pan (2004) report an exposure-response coefficient of 4.8 percent (S.E. $= 0.04$) per $10\,\mu g/m^3$ PM_{10} for bronchitis in adults and 3.4 percent per $10\,\mu g/m^3$ PM_{10} (S.E. $= 0.03$) for bronchitis in children. Altogether, eight cross-sectional questionnaire surveys addressing a range of persistent/chronic respiratory symptoms and diseases were included in Aunan and Pan (2004). All surveys were carried out in Chinese cities, and covered both urban and suburban areas. The coefficients for bronchitis are the result of a meta-analysis of the sub-sample of odds ratios estimated for this particular endpoint (given for Lanzhou, Wuhan, and Benxi). In the studies, the definition of bronchitis was not precise in terms of ICD-9 (or ICD-10) code, but was described as "chronic" or "diagnosed by a physician." We assume that the endpoint approximates chronic bronchitis, and use the relative risk function (1) with $\beta = 0.0048$ for chronic bronchitis in adults.

3.3   Calculating health effects attributable to air pollution

Although the relative risk functions in (1) and (2) can be used to calculate the impact of any change in PM_{10} concentrations, including the benefits of a reduction in PM_{10} achieved by specific pollution control programs, we use them in this chapter to calculate the impact of current ambient PM_{10} levels on health, relative to a threshold level of $15\,\mu g/m^3$. This is the reference level used by the WHO in computing the burden of disease attributed to outdoor air pollution (Cohen et al. 2004).

The number of cases of each health endpoint attributed to air pollution (E) is calculated as the size of the exposed population (P_e) times the difference between the current incidence rate (f_p) and the incidence rate in a clean environment (f_t) [equation (3)]. The latter is calculated from the fact that the current incidence rate equals the "clean" incidence rate times the relative risk, RR. Substituting (4) in (3) implies that (e.g.) excess deaths are the product of current deaths ($f_p P_e$) times the fraction of deaths attributable to air pollution—$(RR - 1)/RR$. Formally,

$$E = (f_p - f_t)P_e \tag{3}$$

$$f_p = f_t^* RR \tag{4}$$

implying

$$E = ((RR - 1)/RR)f_p P_e. \tag{5}$$

The details on incidence data used to compute cases of mortality and morbidity attributable to air pollution appear in the Appendix.

3.4   Cases of premature mortality and morbidity attributed to air pollution

By combining baseline cases of each health endpoint with the selected relative risk functions, we arrive at estimates of the number of excess cases of premature mortality, hospital admissions, and chronic bronchitis attributable to PM_{10}. These estimates are summarized in Table 1.5,

Table 1.5   Health effects associated with outdoor air pollution in China, 2003 (000s)

Estimate	Excess Deaths	Morbidity			In-Hopsital Workdays Lost
		Chronic Bronchitis	Respiratory Hospital Admissions	Cardiovascular Hospital Admissions	
95th %ile	628.3	341.9	286.0	324.3	12,970
Mean	394.0	305.3	223.6	216.3	9,210
5th %ile	134.6	265.6	156.5	99.2	6,108

which presents both mean estimates and the endpoints of a 95 percent confidence interval for each health endpoint. We estimate that approximately 400,000 excess deaths were associated with outdoor air pollution in Chinese cities in 2003, although the endpoints of the confidence interval are wide (135,000 to 600,000). The interval estimates are narrower for new cases of chronic bronchitis and for hospital admissions.

3.4.1   *Premature mortality attributable to air pollution*

Figure 1.4 presents the province-specific estimates of mean excess deaths because of air pollution, indicating both the absolute number of deaths and the percent of urban deaths attributable to air pollution in each province. The results, in general, track the population exposures in Appendix Table A.1.1. Excess deaths as a percentage of total deaths are

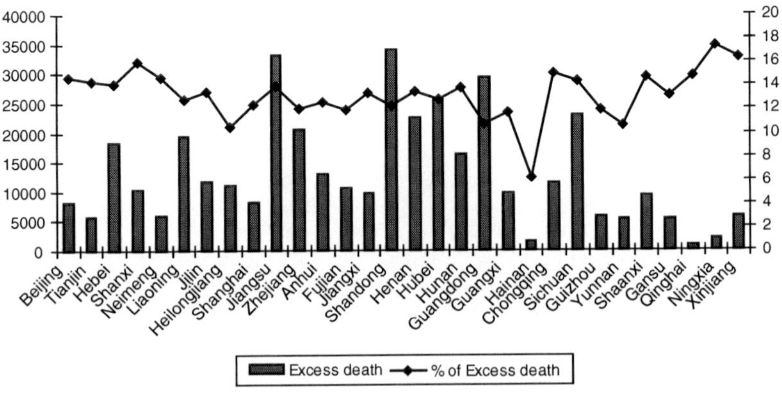

Figure 1.4   Excess deaths attributed to ambient air pollution, by province

generally highest in the provinces in which a high percent of the urban population are in Class III and worse than Class III cities.[11]

As Figure 1.4 makes clear, the provinces with the highest *percentage* of urban deaths because of air pollution are not necessarily the same provinces in which *total deaths* because of air pollution are highest. The provinces with the highest total deaths are Shandong, Jiangsu, and Guangdong, which have large populations as well as high PM_{10} levels. In percentage terms, however, Ningxia and Xinjiang have the highest percent of urban deaths attributed to air pollution, indicating extremely high pollution levels but low urban populations.

3.4.2 Morbidity attributable to air pollution

Table 1.5 indicates that over 300,000 new cases of chronic bronchitis were attributed to ambient air pollution in urban areas of China in 2003, together with almost 450,000 hospital admissions for either respiratory or cardiovascular disease. The number of lost workdays incurred by patients admitted to the hospital or by their families exceeds 9 million. Figure 1.5 maps hospital admissions attributed to air pollution and associated workday losses by province. As Figure 1.5 indicates, the highest numbers of admissions and workday losses attributed to air pollution occur in the eastern and central parts of the country.

4 Monetized health costs

As noted in the Introduction, we monetize the mortality costs of air pollution using two approaches—the Adjusted Human Capital approach (AHC) and an estimate of the Value of a Statistical Life (VSL) based on a stated preference study conducted by Krupnick et al. (2006). These approaches are briefly discussed in this section, which also discusses the approaches used to value morbidity. We then present estimates of the value of avoiding cases of premature mortality and morbidity computed in Section 3.

4.1 Valuing premature mortality

4.1.1 Estimates of the Value of a Statistical Life in China

The approach to valuing premature mortality used in the United States and many other Organization for Economic Co-operation and Development (OECD) countries is to estimate what individuals will pay for small reductions in their risk of dying. The sum of willingnesses to pay (WTP) for risk changes that sum to one statistical life is termed the Value of a

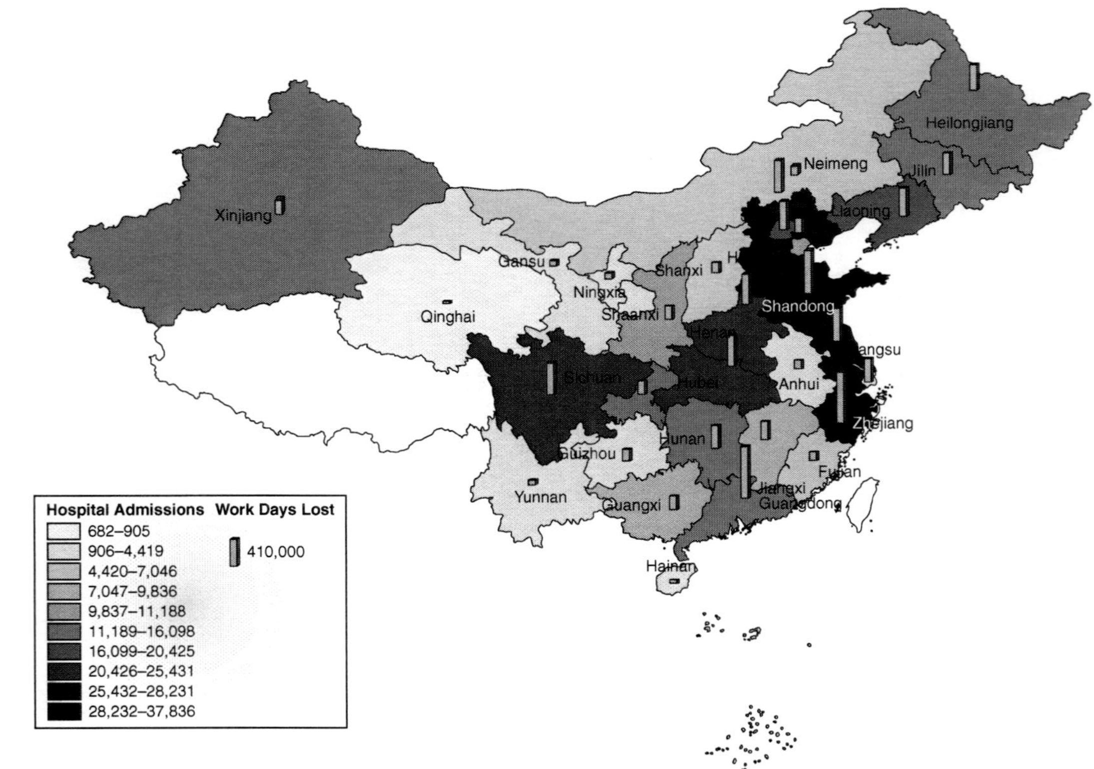

Figure 1.5   Excess hospital admissions and associated workday losses attributed to air pollution, 2003

Statistical Life (VSL). In the United States, estimates of the VSL are based primarily on compensating wage differentials in the labor market (i.e., on hedonic wage studies), although stated preference (e.g., contingent valuation) studies are also used. Although two hedonic wage studies have been conducted in Taiwan (Hammitt and Liu 1999; Liu et al. 1997), similar studies have not been conducted in mainland China.

To our knowledge, four contingent valuation studies have been conducted in China to value quantitative reductions in risk of death: Hammitt and Zhou (2005), Krupnick et al. (2006), Wang and Mullahy (2006), and Zhang (2002). The VSLs obtained in these studies, based on mean WTP, are listed in Table 1.6. VSLs range from 250,000 to 1.7 million RMB Yuan, depending on the study and model used to fit the data. Because contingent valuation studies ask hypothetical questions, it is standard practice for these studies to include tests of internal and external validity of responses. External scope tests vary the size of the risk reduction valued across respondents to see whether WTP increases with the size of the risk reduction. Failure of WTP to increase with the size of the risk reduction suggests that respondents do not perceive risk changes correctly, or are valuing a generalized commodity ("good health") rather than a quantitative risk reduction. Internal scope tests check to see whether WTP increases with the size of the risk reduction for a given respondent. Tests of external validity also include checking whether responses vary, as expected, with income.

Only two of the studies in Table 1.6 performed an external scope test (Hammitt and Zhou 2005 and Krupnick et al. 2006) and only respondents in Krupnick et al. (2006) passed the test. We therefore focus on this study.

Krupnick et al. (2006) conducted contingent valuation surveys in Shanghai and Chongqing in the winter and summer of 2005, respectively, with a second survey in Shanghai in the spring of 2006.

Table 1.6   Estimates of the Value of a Statistical Life in Chinese studies

Study	Million RMB Yuan
Wang and Mullahy (2006)	0.3–1.25
Zhang (2002)	0.24–1.7
Hammitt and Zhou (2005)	0.26–0.51
Krupnick et al. (2006)	1.40

Source: Wang and Mullahy (2006); Zhang (2002); Hammitt and Zhou (2005); Krupnick et al. (2006).

The survey questionnaire, with minor changes, was identical to those administered in the United States, Canada, United Kingdom, France, Italy, and Japan by Alberini et al. (2004). The target population was persons 40- to 80-years-old. Respondents were asked how much they would pay over the next 10 years for a product that would reduce their risk of dying, over the 10-year period, by 10 in 1000 and by 5 in 1000 (i.e., by 10 in 10,000 and 5 in 10,000 per year). Bids were elicited by either a double-bounded dichotomous choice (DC) method or a payment card (PC). The questionnaire was self-administered on a computer with voiceovers.

Samples, stratified by community and neighborhood, were drawn at random in each city. In Shanghai, 1920 persons were initially contacted and invited to take the DC survey, and 1224 participated, an acceptance rate of 64 percent. Another 600 accepted the PC version of the survey. In Chongqing, 1250 persons were contacted and invited to take the survey; 1067 enrolled, a response rate of 85.4 percent.

The results pass some validity tests and not others. The external scope test (in which the WTP for a 5-in-10,000 risk reduction by one group is compared to that of a 10-in-10,000 risk reduction by another group) was passed by the general population using the PC method, but only by highly educated people in Chongqing using the DC method. The regression results are reasonably intuitive and conform to expectations. For instance, those persons with more income, more education, and who are in poorer health are willing to pay more for the risk reduction.

In valuing premature mortality because of air pollution, we use the preferred VSL reported by Krupnick et al. (2006), 1.4 million Yuan, based on pooled data from Shanghai and Chongqing, but adjusted to reflect differences in income between Shanghai, Chongqing, and the rest of China. Once the income adjustment is made, the Krupnick et al. (2006) figure is approximately 1 million Yuan.[12] We note that this falls within the range of values reported in the other studies listed in Table 1.6. Following the practice used in the United States and Europe, we apply the same value to all lives lost because of air pollution, regardless of location (i.e., of per capita GDP). This practice is followed in the United States for political, rather than economic, reasons.

4.1.2   *The adjusted human capital approach*

An alternate approach to WTP is to use the productivity loss associated with premature mortality (i.e., forgone earnings) to value loss of life. This values an individual by what he produces and assumes that this value is accurately measured by his earnings. The adjusted human capital (AHC) approach, which is widely used in China, represents an important departure from the traditional human capital approach. Because the use of foregone earnings would assign a value of zero to the lives of the retired and the disabled, the AHC approach avoids this problem by assigning the same value—per capita GDP—to a year of life lost by all persons, regardless of age. For this reason, the AHC approach can be viewed as a social statement of the value of avoiding premature mortality.

In practice, the AHC values a life lost at any age by the present discounted value of per capita GDP over the remainder of the individual's expected life. In computing the AHC measure, real per capita GDP is assumed to grow at rate a annually and is discounted to the present at the rate r. Adjusted human capital, HC_m, is thus given by (6)

$$HC_m = GDP_{pc0} \sum_{i=1}^{t} \frac{(1+\alpha)^i}{(1+r)^i} \qquad (6)$$

where GDP_{pc0} is per capita GDP in the base year and t is remaining life expectancy. In the base case calculations $\alpha = 7\%$ and $r = 8\%$, which are official government values.

Equation (6) implies that HC_m will vary with the age of the person who dies and will vary by city or province, assuming that per capita GDP varies by city or province. Remaining life expectancy, which does not vary by province in the published data, is calculated using Chinese life tables assuming that the age distribution of deaths because of air pollution is identical to the age distribution of deaths because of respiratory and cardiovascular diseases. As shown in Appendix Table A.1.2, the average number of life-years lost because of air pollution is approximately 18. Per capita GDP in the base year (2003) differs by city. Table 1.7 shows the AHC measure computed for different cities, assuming $r = 8\%$ and allowing a to equal 6%, 7%, and 8%. The central case estimates below correspond to HC_m in the second column from the right.[13]

4.2 Valuing morbidity

In principle, economists value avoided morbidity by the amount a person will pay to avoid (the risk of) an illness, just as risk of death is valued by what people will pay to reduce it. In the case of morbidity, WTP should capture the value of the pain and suffering avoided, as well as the value of time lost because of illness (both leisure and work time) and the costs of medical treatment. If some of these costs are not borne

Table 1.7 Adjusted Human Capital (HC_m) measure for different cities with different growth rates of per capita GDP (Base year: 2003)

Growth rate of GDP/capita (α, %)		6	7	8
Discount rate (r, %)		8	8	8
$\sum_{i=1}^{t}[(1+\alpha)/(1+r)]^i$	Value	15.14	16.50	18
Cities	Per Capita GDP (Yuan)	HC_m (10,000 Yuan)		
Beijing	32,061	48.55	52.89	57.71
Tianjin	26,532	40.18	43.77	47.76
Shijiazhuang	15,188	23.00	25.06	27.34
Taiyuan	15,210	23.03	25.09	27.38
Huhehaote	18,791	28.45	31.00	33.82
Shenyang	23,271	35.24	38.39	41.89
Dalian	29,206	44.22	48.18	52.57
Changchun	18,705	28.32	30.86	33.67
Haerbin	14,872	22.52	24.53	26.77
Shanghai	46,718	70.74	77.07	84.09
Nanjing	27,307	41.35	45.05	49.15
Hangzhou	32,819	49.70	54.14	59.07
Ningbo	32,639	49.42	53.84	58.75
Hefei	10,720	16.23	17.68	19.30
Fuzhou	20,520	31.07	33.85	36.94
Xiamen	35,009	53.01	57.75	63.02
Nanchang	14,382	21.78	23.73	25.89
Jinan	23,590	35.72	38.92	42.46
Qingdao	23,398	35.43	38.60	42.12
Zhengzhou	17,063	25.84	28.15	30.71
Wuhan	21,457	32.49	35.40	38.62
Changsha	14,810	22.43	24.43	26.66
Guangzhou	48,372	73.25	79.80	87.07
Shenzhen	54,545	82.59	89.98	98.18
Nanning	7,874	11.92	12.99	14.17
Haikou	16,730	25.33	27.60	30.11
Chongqing	8,077	12.23	13.32	14.54
Chengdu	18,051	27.33	29.78	32.49
Guiyang	10,962	16.60	18.08	19.73
Kunming	16,312	24.70	26.91	29.36
Xian	12,233	18.52	20.18	22.02
Lanzhou	14,540	22.02	23.99	26.17
Xining	7,110	10.77	11.73	12.80
Yinchuan	11,788	17.85	19.45	21.22
Wulumuqi	19,900	30.13	32.83	35.82

Source: Authors' calculations.

by the individual, and are therefore not reflected in his willingness to pay, the value of the avoided costs must be added to WTP to measure the social benefits of reduced morbidity.

In cases where WTP estimates are not available, analysts often rely on cost-of-illness (COI) estimates as a lower bound to the theoretically correct value of avoiding illness. COI studies estimate the lost earnings associated with chronic illness that result from both reduced labor force participation and lower earnings conditional on participation (Bartel and Taubman 1979; Krupnick and Cropper 2000), and add to these medical costs associated with the disease. The COI is a lower bound to WTP because it ignores the value of pain and suffering associated with illness and the value of lost leisure time. In regulatory impact analyses of air pollution regulations published by the U.S. Environmental Protection Agency (USEPA 1997); it is often the case that coronary heart disease and stroke are valued using COI estimates, as WTP estimates are unavailable.

In this chapter, we approximate WTP for chronic bronchitis using benefits-transfer methods. For hospital admissions, we rely on COI estimates.

4.2.1 *Valuing chronic bronchitis*

In the case of common illnesses, such as diarrheal disease, economists usually value reductions in days of illness, treated as certain. For illnesses that are rarer, such as chronic bronchitis, it is appropriate to view exposure to pollutants as increasing the risk of serious illness and to value reductions in risk of illness.

To value reductions in the risk of chronic bronchitis, one could ask individuals directly what they would pay to lower their risk of experiencing these conditions. An alternate approach that has proved successful (Viscusi et al. 1991) is to ask individuals to make trade-offs between the risk of contracting a serious illness and the risk of death (e.g., dying in an auto accident). These risk-risk trade-offs establish an equivalence between the utility of good health and the utility of the disease. For example, in a U.S. study involving trade-offs between risk of contracting chronic bronchitis and risk of dying in an auto accident, people's choices implied that the utility of living with chronic bronchitis was about 0.68 of the utility of living in good health (Viscusi et al. 1991). If good health is scaled to equal 1 and death is scaled to equal 0, then this is equivalent to saying that living a year with chronic bronchitis is equal to losing 0.32 of a year of life. This number can be converted to the value of a statistical case of chronic bronchitis by multiplying the VSL by 0.32.

The risk-risk trade-off approach is closely related to methods used in the public health literature to establish QALY weights for chronic disease—the ratio of the utility of living with the disease to the utility of living in good health (Miller et al. 2006).[14] It is, therefore, possible to draw on the QALY literature to establish the fraction of a year lost if one has chronic bronchitis. Clearly, this equivalence will depend on the severity of the case of chronic bronchitis. It is, therefore, not surprising that the QALY weights reported in the literature for chronic bronchitis vary widely.

Although one attempt has been made to estimate a QALY weight for chronic bronchitis in China, we choose a value from the international literature. In survey work in China, Hammitt and Zhou (2005) use both risk-risk trade-offs and standard gambles to determine the utility lost because of chronic bronchitis. However, the case of chronic bronchitis they describe is a very mild one. We, therefore, appeal to the international literature on QALY weights for chronic bronchitis, and select a value in the middle of the range of weights reported by Miller et al. (2006, Appendix A). Specifically, we assume that living a year with chronic bronchitis is equivalent to losing 0.4 years of life.

When excess deaths are valued using the VSL from Krupnick et al. (2006), the value of a statistical case of chronic bronchitis is computed as 0.4*VSL. When the AHC approach is used to value excess deaths, we compute HC_m using the expected number of years a person will live with chronic bronchitis in place of t in equation (6) and multiply the result by 0.4.

4.2.2   Valuing hospital admissions

For most acute illness episodes (restricted activity days, asthma attacks), contingent valuation is the method most often used to value avoided morbidity (Freeman 1993; Loehman and De 1982). In China, few contingent valuation studies have been conducted to value acute illness. Notable exceptions are Hammitt and Zhou (2005), who estimate WTP to avoid a cold in Anqing and Beijing, and studies conducted in Taiwan to estimate WTP to avoid a recurrence of acute respiratory illness (Alberini et al. 1997). Unfortunately, we know of no studies in China that estimate WTP to avoid a respiratory or cardiovascular hospital admission. We, therefore, use the COI approach to value hospital admissions.

National surveys on health services were carried out in China in 1998 and 2003 in which medical costs were reported. The 1998 survey

Table 1.8   Illness costs for hospital admissions in China in 2003 (Yuan/episode)

Cause of Admission	Direct plus Indirect Costs			Indirect Cost
	Large-scale city	Middle-scale city	Small-scale city	
Respiratory	8,474	5,071	2,593	514
Cardiovascular	12,326	8,506	6,028	514

Source: Authors' calculations based on the China National Health Survey 2003.

provided disease-specific medical cost information, whereas the 2003 survey provided only all-disease average costs. However, the 2003 report calculated the increase in average medical cost from 1998 to 2003. Assuming that each disease-specific cost increased by the same proportion, we estimate the disease-specific costs in 2003, as shown in Table 1.8. The direct costs of illness include all the costs in hospital, including expenditures for medical examinations, drugs, and therapy, as well as the cost of the hospital stay. Indirect costs include the patient's time lost from work, as well as the workdays lost by patients' families. In China, it is common that the family, colleagues, or friends of the patients leave their work to visit the patients in hospital. The economic loss from this kind of work absence has been valued as well. Illness costs are broken down by city size, as well as type of hospital admission.

4.3   Monetary health costs of ambient air pollution

Tables 1.9 and 1.10 summarize the monetary costs of ambient air pollution. Table 1.9 summarizes the costs of ambient air pollution using

Table 1.9   Health costs associated with outdoor air pollution in China, 2003 Adjusted Human Capital Approach (Bil. Yuan)

Estimate	Excess Deaths	Morbidity			Total Costs
		Chronic Bronchitis	Direct Hospital Costs	Indirect Hospital Costs	
95th %ile	178.7	47.7	4.82	0.670	231.8
Mean	110.9	42.5	3.41	0.470	157.3
5th %ile	35.8	36.9	1.88	0.264	74.9

Source: Authors' calculations.

Table 1.10   Health costs associated with outdoor air pollution in China, 2003 Willingness to pay approach (Bil. Yuan)

Estimate	Excess Deaths	Morbidity			Total Costs
		Chronic Bronchitis	Direct Hospital Costs	Indirect Hospital Costs	
95th %ile	641.1	136.7	4.82	0.670	783.3
Mean	394.0	122.1	3.41	0.470	519.9
5th %ile	135.6	106.2	1.88	0.263	243.9

Source: Authors' calculations.

the AHC approach to value both premature mortality and chronic bronchitis. Table 1.10 repeats the calculations using the VSL to monetize premature mortality and chronic bronchitis. The mean estimates and 5th and 95th percentiles refer to the uncertainty bounds for the number of cases of mortality and morbidity.

Several points are worth noting. The mean total health cost associated with ambient air pollution in urban areas of China in 2003 is 157 billion Yuan if the AHC approach to valuation is used, and 520 billion if WTP estimates from Krupnick et al. (2006) are used. Use of WTP increases total costs by a factor of 3.3, bringing health costs to 3.8 percent of 2003 GDP. Using the AHC approach, health costs are 1.2 percent of GDP. As in many studies, the damages associated with premature mortality dominate the total: they are 71 percent of health costs using the AHC approach and 76 percent using the WTP approach. However, in both cases chronic bronchitis costs are significant—over 20 percent of total costs.

Health costs are broken out by province in Figures 1.6 and 1.7. Total health costs, when valued using the AHC approach, reflect total cases of illness and premature mortality, as well the cost per case avoided, which, for premature mortality and chronic bronchitis, is proportional to urban per capita GDP.[15] If one examines total costs by province, Jiangsu, Guangdong, and Shandong have the highest total costs, followed by Zhejiang, Shanghai, and Liaoning (see Figure 1.6). Jiangsu, Guangdong, and Shandong are the provinces with the highest number of excess deaths and cases of chronic bronchitis because of air pollution. They also have above average costs per case. Shanghai, on the other hand, has only one-fourth of the excess deaths of Shandong, but has the highest AHC cost per death.

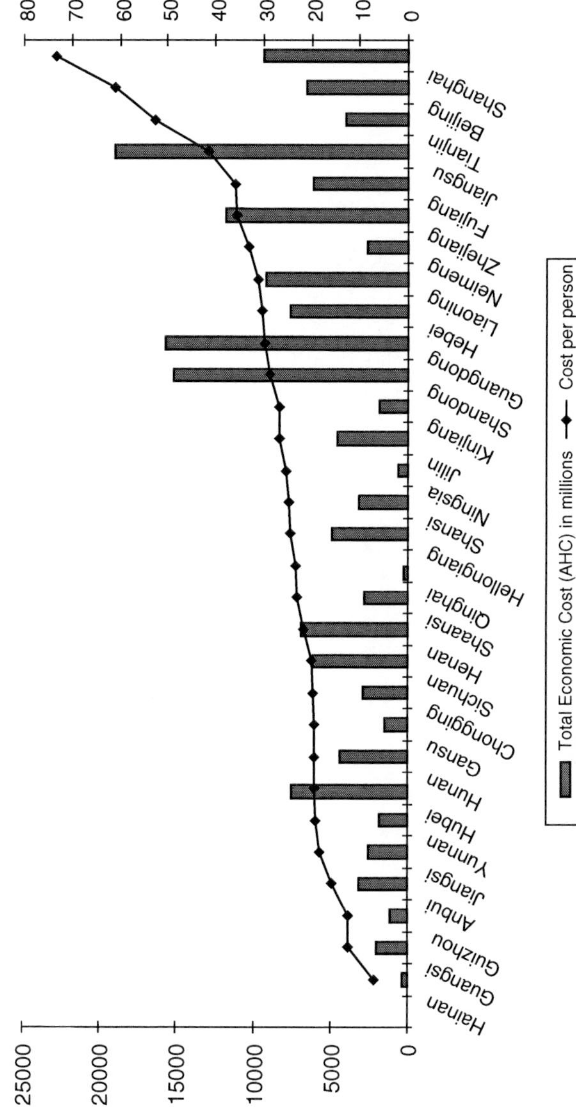

Figure 1.6 Total health cost and per capita health cost due to air pollution
Source: Authors' calculations.

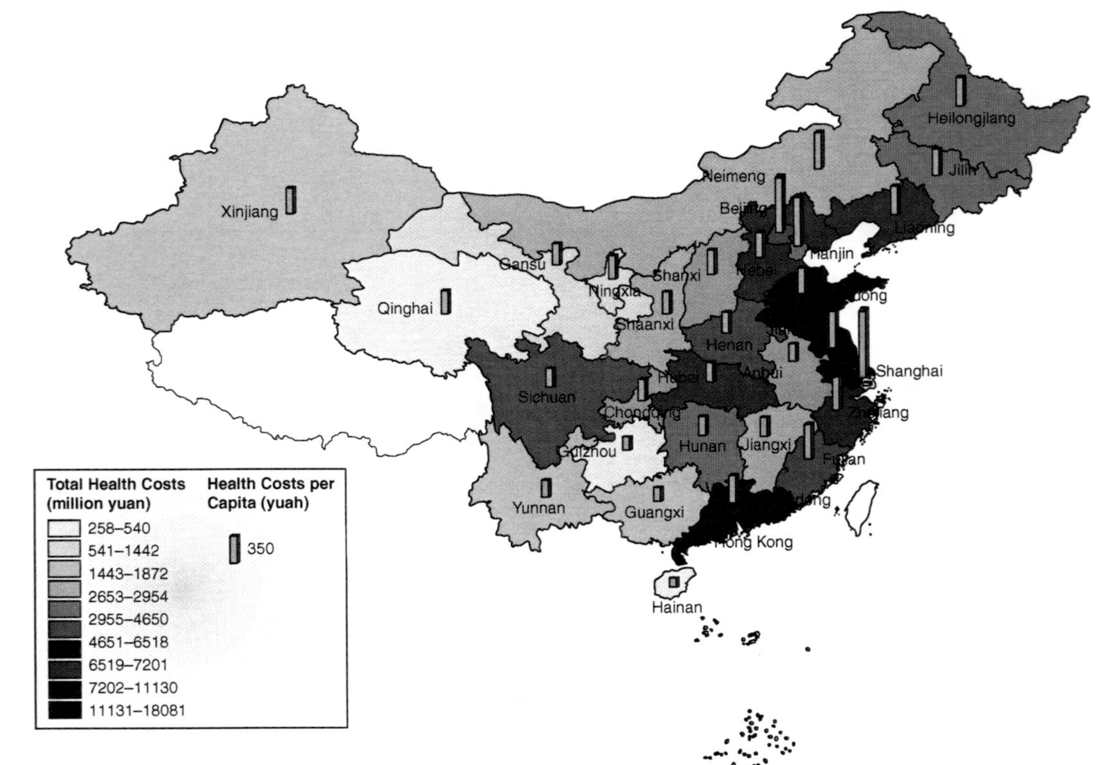

Figure 1.7   Health costs associated with outdoor air pollution, 2003
Source: Authors' calculations.

The ranking of provinces is quite different based on AHC costs per capita. Figure 1.6 sorts provinces by the AHC health cost of air pollution per person. AHC costs per person are proportional to ambient PM_{10} concentrations (which increases cases of illness and premature mortality) and to per capita GDP, but do not depend on population. The three provinces/municipalities with the highest health costs on a per capita basis are Shanghai, Beijing, and Tianjin, which have the highest per capita GDP and hence the highest AHC costs per premature death and per case of chronic bronchitis.

Finally, the ranking of provinces based on health costs as a percent of urban GDP reflects only population-weighted PM_{10} concentrations, not total exposed population or per capita GDP. Shanxi, Xinjiang, Neimeng, and Ningxia all have health damages in excess of 2 percent of urban GDP. This reflects the high number of cases of premature mortality and chronic bronchitis per person in these provinces, a result foreshadowed by Figure 1.1.

Which of the measures of health damages is most relevant? From the perspective of evaluating the net benefits of pollution control strategies, the provinces with the highest total health damages associated with outdoor air pollution are likely to have the highest net benefits associated with a given reduction in ambient PM_{10} concentrations, as long as the per person costs of the strategy are roughly the same in all provinces. From an equity perspective, high pollution costs expressed as a percent of GDP suggest a high per capita physical burden associated with air pollution. This may be a relevant consideration to policy makers, as may high per capita monetary costs, which primarily reflect high per capita GDP.

5 Conclusion

It would be a mistake to set environmental priorities based on the total damages associated with air pollution, just as it would be a mistake to set health priorities based on WHO's Global Burden of Disease. Which pollution control measures should be adopted in a country and where they should be adopted should depend in part on a comparison of benefits and costs. The analyses presented here are nonetheless interesting in presenting a picture of the damages associated with air pollution in China and their geographic distribution. More importantly, the methods presented here can be used to compute the benefits of pollution control measures for use in benefit cost analyses.

Appendix

Calculation of baseline incidence (f_p)

Hospital admissions

The *Health Statistical Yearbook* (Ministry of Health 2004a) provides only all-cause hospital admissions by province, and not hospital admissions for specific diseases such as respiratory disease. Another problem is that hospital admissions by province include both rural and urban areas, whereas only the urban population is used to calculate the health costs of air pollution. We estimate hospital admissions for respiratory disease in urban areas in two steps. First, we estimate the number of hospital admissions because of respiratory diseases by multiplying all-cause hospital admissions by the ratio of respiratory diseases to all diseases by province. The percentage of respiratory disease to all diseases is reported in the Analysis Report of the *Third National Health Services Survey* (Ministry of Health 2004b). This is based on an assumption that the share of patients being admitted to the hospital for respiratory diseases resembles the share of people suffering from respiratory diseases among all people who are ill. Second, we estimate the number of hospital admissions because of respiratory disease in the urban population from the ratio of the urban population to the total population. This is based on an assumption that the hospitalization rate per case of disease is the same in urban and rural areas, which is a crude approximation.

Premature mortality

Current mortality rates, which vary by city size, are obtained from the *China Health Statistical Yearbook*.

Chronic bronchitis

Calculating annual cases of chronic bronchitis associated with air pollution requires an estimate of the incidence of chronic bronchitis by city. Because only prevalence rates are available, we approximate the annual incidence of chronic bronchitis by dividing the prevalence rate by the average duration of the illness (23 years). This yields an incidence rate of approximately 0.00148.

Table A.1.1 PM$_{10}$ pollution exposure of the urban population (population in 10,000's)

Provinces	Item	Class I	Class II	Class III	> Class III	Total
		PM$_{10}$ < $40\,\mu g/m^3$	PM$_{10}$: $40-100\,\mu g/m^3$	PM$_{10}$: $100-150\,\mu g/m^3$	PM$_{10}$ > $150\,\mu g/m^3$	Population/%
Beijing	Population	0	0	1,079	0	1,079
	%	0.00	0.00	100.00	0.00	100
Tianjin	Population	0	0	759	0	759
	%	0.00	0.00	100.00	0.00	100
Hebei	Population	0	384	1,650	496	2,529
	%	0.00	15.16	65.23	19.60	100
Shanxi	Population	0	148	322	796	1,267
	%	0.00	11.68	25.45	62.87	100
Neimeng	Population	0	144	311	240	694
	%	0.00	20.67	44.74	34.59	100
Liaoning	Population	0	1,615	1,265	78	2,958
	%	0.00	54.61	42.75	2.64	100
Jilin	Population	0	807	473	407	1,687
	%	0.00	47.80	28.06	24.14	100
Heilong Jiang	Population	0	627	841	179	1,647
	%	0.00	38.08	51.04	10.88	100
Shanghai	Population	0	1,278	0	0	1,278
	%	0.00	100.00	0.00	0.00	100
Jiangsu	Population	0	639	3516	458	4,613
	%	0.00	13.84	76.22	9.94	100
Zhejiang	Population	0	1,532	1,782	0	3,314
	%	0.00	46.22	53.78	0.00	100

Table A.1.1 (Continued)

Provinces	Item	Class I	Class II	Class III	> Class III	Total Population/%
		$PM_{10} <$ $40\,\mu g/m^3$	PM_{10}: $40 - 100\,\mu g/m^3$	PM_{10}: $100 - 150\,\mu g/m^3$	PM_{10} $> 150\,\mu g/m^3$	
Anhui	Population	0	927	1062	0	1,990
	%	0.00	46.61	53.39	0.00	100
Fujian	Population	0	1,243	385	79	1,707
	%	0.00	72.81	22.57	4.62	100
Jiangxi	Population	0	448	800	159	1,407
	%	0.00	31.85	56.85	11.30	100
Shandong	Population	0	3,610	1,546	190	5,345
	%	0.00	67.53	28.92	3.55	100
Henan	Population	0	792	1,706	738	3,236
	%	0.00	24.47	52.72	22.81	100
Hubei	Population	0	1,520	2,351	0	3,871
	%	0.00	39.26	60.74	0.00	100
Hunan	Population	0	317	1,594	358	2,269
	%	0.00	13.97	70.23	15.80	100
Guang Dong	Population	0	5,005	293	0	5,298
	%	0.00	94.47	5.53	0.00	100
Guangxi	Population	204	473	689	254	1,619
	%	12.57	29.20	42.55	15.68	100
Hainan	Population	354	113	0	0	467
	%	75.86	24.14	0.00	0.00	100
Chong Qing	Population	0	0	1,488	0	1,488
	%	0.00	0.00	100.00	0.00	100

Sichuan	Population	0	489	1,337	1,276	3,103
	%	0.00	15.77	43.09	41.14	100
Guizhou	Population	0	357	582	0	939
	%	0.00	38.00	62.00	0.00	100
Yunnan	Population	64	789	76	0	929
	%	6.91	84.95	8.14	0.00	100
Xizang	Population	0	14	0	0	14
	%	0.00	100.00	0.00	0.00	100
Shaanxi	Population	0	147	721	340	1,207
	%	0.00	12.19	59.69	28.13	100
Gansu	Population	0	124	339	272	735
	%	0.00	16.92	46.04	37.04	100
Qinghai	Population	0	0	107	12	119
	%	0.00	0.00	89.92	10.08	100
Ningxia	Population	0	0	72	157	229
	%	0.00	0.00	31.34	68.66	100
Xinjiang	Population	0	181	278	226	684
	%	0.00	26.37	40.66	32.96	100
Total	Population	622	23,720	27,422	6,716	58,480
	%	1.06	40.56	46.89	11.48	100

Note: The PM$_{10}$ pollution exposure is computed based on data from 660 cities. See note 3.
Source: Authors' calculations.

Table A.1.2 Average life years lost due to respiratory and cardiovascular diseases

Age groups	Remaining Life Expectancy	RD		CVD		CEVD	
		Deaths	Lost life years × Deaths	Deaths	Lost life years × Deaths	Deaths	Lost life years × Deaths
0–	78.79	1680.41	132393.79	266.69	21011.27	89.25	7031.41
1–4	78.51	518.07	40675.18	130.56	10250.80	21.93	1722.13
5–9	74.71	237.68	17756.99	68.13	5089.58	9.08	678.61
10–14	69.83	138.12	9644.70	138.12	9644.70	30.47	2127.51
15–19	64.92	195.99	12723.77	229.12	14874.26	99.38	6451.49
20–24	60.01	253.55	15215.41	548.45	32911.59	212.21	12734.64
25–29	55.15	480.56	26502.25	964.25	53176.60	436.88	24092.96
30–34	50.31	987.78	49696.39	2019.46	101601.51	1153.98	58058.01
35–39	45.50	1274.53	57991.06	3279.82	149232.08	2359.17	107342.37
40–44	40.73	1797.80	73231.00	4369.61	177990.75	4169.36	169833.60
45–49	36.07	3519.17	126930.67	7359.60	265448.36	8821.92	318191.93
50–54	31.49	4903.51	154393.74	8674.55	273130.10	10787.44	339657.12

55–59	27.06	6598.43	178537.66	9180.92	248413.60	12547.36	339501.39
60–64	22.86	14205.99	324757.07	16643.19	380473.08	21818.92	498793.14
65–69	19.01	25778.30	489933.95	29385.85	558497.67	36225.79	688495.36
70–74	15.64	41228.53	644924.21	39312.80	614957.14	47827.45	748148.97
75–79	12.96	46403.88	601328.17	40830.47	529104.65	48111.39	623455.11
80–85	11.07	41399.67	458171.04	34687.21	383884.08	36150.15	400074.47
85–	10.72	36785.43	394314.94	31578.30	338497.99	25073.53	268771.30
Total		228387.42	3809121.98	229667.09	4168189.83	255945.66	4615161.51
Average lost statistical years			16.68		18.15		18.03

Note: Deaths are the product of the population in the survey report of the national 5th population census and the disease-specific death rates in the *Health Statistical Yearbook*.
RD = Respiratory disease; CVD = Cardiovascular disease; CEVD = Cerebrovascular disease.

Notes

*This chapter was coauthored with Kristin Aunan, CICERO, Oslo, Norway, and Pan Xiaochuan and Zhang Yanshen of the Public Health School of Peking University, Beijing, 100083, China. The findings, interpretations, and conclusions expressed in this chapter are entirely those of the authors. They do not necessarily represent the views of the institutions with which they are affiliated.

1. In 2000, approximately 44 percent of China's coal was burned by electric utilities, half by industry and the remainder by households.
2. See, for example, *The Benefits and Costs of the Clean Air Act, 1990 to 2010* (USEPA, 1999).
3. Currently, air quality is monitored in approximately 340 Chinese cities. In 2003, PM_{10} was monitored in 228 of 341 cities. Estimates of PM_{10} were computed for the remaining 113 cities based on readings of total suspended particulates (TSP). Specifically, the ratio of PM_{10}/TSP was computed for each province, based on the cities in which both pollutants were monitored, and applied to TSP readings in cities where only TSP was monitored. Pollution levels in non-monitored county-level cities were estimated based on data in their upper-level prefecture cities.
4. See, for example, USEPA (1997, 1999) and Miller et al. (2006).
5. We thank Rick Burnett for performing these calculations.
6. The Dockery et al. (1993) and Pope et al. (1995, 2002) studies are similar in some respects to the ecologic cross-sectional studies because the variation in pollution exposure is measured across locations rather than over time. They rely on the same type of pollutant exposure data as that used in the ecologic studies, which is based on average pollutant levels measured at stationary outdoor monitors in a given location. However, these studies use individual-level data so that other health risk factors can be better characterized. Specifically, the authors of the prospective studies are able to control mortality risks associated with differences in body mass, occupational exposures, smoking (present and past), alcohol use, age, and gender.
7. Specifically, we apply a conversion ratio of 0.60 for $PM_{2.5}$ to PM_{10} and a ratio of 0.50 for PM_{10} to TSP. Aunan and Pan (2004) suggest that the conversion ratio of TSP to PM_{10} is 0.60. In Dockery's six-city study (Dockery et al. 1993), the ratio of $PM_{2.5}$ to PM_{10} is 0.60 to 0.64. In the recent Chinese four-city study (Qian et al. 2001), the ratio is $0.51 \sim 0.72$. Wan (2005) found an average ratio of 0.55 in 28 cities in China.
8. It should be noted that Pope et al. (1995, 2002) find no significant relationship between PM_{10} and mortality, only between $PM_{2.5}$ and mortality. In constructing Table 1.4, we assume that the Beta coefficients in Pope et al. apply to the portion of PM_{10} that constitutes $PM_{2.5}$. Assuming that $\Delta PM_{2.5} = 0.6 \Delta PM_{10}$ and because (1) is approximately linear in ΔC, we multiply the Beta coefficients in Pope et al. by 0.6.
9. Personal communication from Rick Burnett, July 2006.
10. Ideally, we would like to estimate γ in equation (2) separately for each cause of death and apply the U.S. results to China by cause of death. This is, however, impossible since estimates of death rates by cause are not available for

individual classes of cities. For adults (the population to which (2) will be applied), a comparison of death rates by cause in the United States (Miniño et al. 2006) and China (He et al. 2005) suggests that the distributions of death by cause are close: Deaths attributable to cancer are approximately 23 percent of total deaths in both countries. The percent of deaths because of cardiovascular disease is 27 percent in the United States and 23 percent in China, although cerebrovascular deaths are relatively more important in China (21 percent) than in the United States (6 percent).

11. Note, however, that the shape of the relative risk function in Figure 1.3 implies that a given microgram change in PM_{10} has a larger impact on mortality between 50 and 100 micrograms than between 100 and 150 micrograms. This explains why Guangdong has such a large number of deaths attributed to air pollution.

12. This adjustment is made using the ratio of average disposable income in China to average disposable income in Shanghai and Chongqing. The income elasticity of 0.48 from Krupnick et al. (2006) is used to make the adjustment.

13. It should be noted that the adjusted human capital values in Table 1.7 pertain to cities, whereas the results below are reported for provinces and municipalities.

14. One such approach is the standard gamble approach, used by Hammitt and Zhou (2005). This approach asks a person, were he to contract chronic bronchitis, what risk of death? he would accept to undergo an operation that would cure the disease with probability $1 - \rho$.

15. Total Health Costs = Cases of Premature Mortality*Cost per Case + Cases of Chronic Bronchitis*Cost per Case + Direct Cost of Hospital Admissions + Indirect Cost of Hospital Admissions.

References

Alberini, A., M.L. Cropper, T. Fu, A. Krupnick, J.-T. Liu, D. Shaw, and W. Harrington (1997) 'Valuing Health Effects of Air Pollution in Developing Countries: The Case of Taiwan', *Journal of Environmental Economics and Management,* vol. 34, 107–26.

Alberini, A., M.L. Cropper, A. Krupnick, and N. Simon (2004) 'Does the Value of Statistical Life Vary with Age and Health Status? Evidence from the U.S. and Canada', *Journal of Environmental Economics and Management,* vol. 48, 769–92.

Aunan, K. and X.C. Pan (2004) 'Exposure-Response Functions for Health Effects of Ambient Air Pollution Applicable for China—A Meta-analysis', *Science of the Total Environment,* vol. 329(1–3), 3–16.

Bartel, A. and P. Taubman (1979) 'Health and Labor Market Success: The Role of Various Diseases', *Review of Economics and Statistics,* vol. 61, 1–8.

Chang, G., X. Pan, X. Xie, and Y. Gao (2003) 'Time-series Analysis on the Relationship between Air Pollution and Daily Mortality in Beijing', *Journal of Hygiene Research,* vol. 32(6), 567–69.

Cohen, A.J., H.R. Anderson, B. Ostro, K.D. Pandey, M. Krzyzanowski, N. Kuenzli, K. Gutschmidt, C.A. Pope, I. Romieu, J. Samet, and K. Smith (2004) 'Urban Air Pollution' in M. Ezzati, A.D. Rodgers, A.D. Lopez, and C.J.L. Murray (eds)

Comparative Quantification of Health Risks: Global and Regional Burden of Disease Due to Selected Major Risk Factors, Vol. 2 (Geneva: World Health Organization).

Dockery, D.W., C.A. Pope, X. Xu, J.D. Spengler, J.H. Ware, M.E. Fay, B.G. Ferris, and F.E. Speizer (1993) 'An Association between Air Pollution and Mortality in Six U.S. Cities', *New England Journal of Medicine*, vol. 329(24), 1753–59.

Dong, J., Y. Chen, D.W. Dockery, and J. Jiang (1995) 'Association between Air Pollution and Daily Mortality in Urban Areas in Beijing in 1990–1991', *Journal of Hygiene Research*, vol. 24(4), 212–14.

Gao, J., X. Xu, and Y. Chen (1993) 'Association between Air Pollution and Mortality in Dongcheng and Xicheng Districts in Beijing', *Chinese Journal of Preventive Medicine*, vol. 27(6), 340–43.

Freeman, A.M. III. (1993) *The Measurement of Environmental and Resource Values: Theory and Methods* (Washington, D.C.: Resources for the Future).

Hammitt, J.K. and Liu, Jin-Tan (1999) 'Perceived Risk and Value of Workplace Safety in a Developing Country', *Journal of Risk Research*, vol. 2(3), 263–75.

Hammitt, J.K. and Y. Zhou (2005) 'The Economic Value of Air-Pollution-Related Health Risks in China: A Contingent Valuation Study', *Environmental and Resource Economics*, vol. 33, 399–423.

He, Jiang, D. Gu, X. Wu, K. Reynolds, X. Duan, C. Yao, J. Wang, C.S. Chen, J. Chen, R.P. Wildman, M.J. Klag, and P.K. Whelton (2005) 'Major Causes of Death among Men and Women in China', *New England Journal of Medicine*, vol. 353(1), 124–34.

HEI (2004) *Health Effects of Outdoor Air Pollution in Developing Countries of Asia: A Literature Review.* Health Effects Institute International Scientific Oversight Committee, Special Report 15.

Jing, L., S. Wang, Y. Qin, C. Ren, Z. Xu, L. Ren, J. Li, F. Sha, B. Chen, and T. Kjellstrom (1999) 'Association between Air Pollution and Mortality in Benxi City', *Chinese Journal of Public Health*, vol. 15(3), 211–12.

Kan, H. and B. Chen (2003) 'Air Pollution and Daily Mortality in Shanghai: A Time-series Study', *Archives of Environmental Health*, vol. 58(6), 360–67.

Kan, H., J. Jia, and B. Chen (2004) 'The Association of Daily Diabetes Mortality and Outdoor Air Pollution in Shanghai, China', *Journal of Hygiene Research*, vol. 67(3), 21–25.

Krupnick, A. and M. Cropper (2000) 'The Social Costs of Chronic Heart and Lung Disease' in *Valuing Environmental Benefits, Selected Essays of Maureen Cropper* (Cheltenham, UK: Edward Elgar).

Krupnick, A., S. Hoffmann, B. Larsen, X. Peng, R. Tao, and C. Yan (2006) *The Willingness to Pay for Mortality Risk Reductions in Shanghai and Chongqing, China* (Washington, D.C.: Resources for the Future).

Liu, J.-T., J.K. Hammitt, and J-L. Liu (1997) 'Estimated Hedonic Wage Function and Value of Life in a Developing Country', *Economics Letters*, vol. 57, 353–58.

Loehman, E. and V. De (1982) 'Application of Stochastic Choice Modeling to Policy Analysis of Public Goods: A Case Study of Air Quality Improvements', *Review of Economics and Statistics*, vol. 64, 474–80.

Miller, W., L.A. Robinson, and R.S. Lawrence (eds) (2006) *Valuing Health for Regulatory Cost-Effectiveness Analysis* (Washington, D.C.: The National Academies Press).

Miniño, A.M., M.P. Heron, and B. Smith (2006) 'Deaths: Preliminary Data for 2004', *National Vital Statistics Reports*, vol. 54(19), 1–50.

Ministry of Health, P.R. China (2004a) *China Health Statistical Yearbook 2004* (Beijing: Chinese Academy of Medical Sciences Press).

Ministry of Health, Statistical Information Center, P.R. China (2004b) *Analysis Report of National Health Services Survey 2003* (Beijing: Chinese Academy of Medical Sciences Press).

Ostro, B. (2004) 'Outdoor Air Pollution: Assessing the Environmental Burden of Disease at National and Local Levels'. Geneva, World Health Organization (WHO Environmental Burden of Disease Series, No. 5).

Pandey, Kiran (2006) Personal Communication, March 2006.

Pope, C.A., M.J. Thun, M.M. Namboodiri, D.W. Dockery, J.S. Evans, F.E. Speizer, and C.W. Heath (1995) 'Particulate Air Pollution as a Predictor of Mortality in a Prospective Study of U.S. Adults', *American Journal of Respiratory and Critical Care Medicine*, vol. 151, 669–74.

Pope, C.A., R.T. Burnett, M.J. Thun, E.E. Calle, D. Krewski, K. Ito, and G.D Thurston (2002) 'Lung Cancer, Cardiopulmonary Mortality, and Long-term Exposure to fine Particulate Air Pollution', *Journal of the American Medical Association*, vol. 287(9), 1132–41.

Qian, Z., J. Zhang, F. Wei, W.E. Wilson, and R.S. Chapman (2001) 'Long-term Ambient Air Pollution Levels in Four Chinese Cities: Inter-City and Intra-City Concentration Gradients for Epidemiological Studies', *Journal of Exposure Analysis and Environmental Epidemiology*, vol. 11(5), 341–51.

Samet, J.M., F. Dominici, F.C. Curriero, I. Coursac, and S.L. Zeger (2000) 'Fine Particulate Air Pollution and Mortality in 20 U.S. Cities, 1987–1994', *New England Journal of Medicine*, vol. 343(24), 1742–49.

Schwartz, J., C. Spix, G. Touloumi, L. Bacharova, T. Barumamdzadeh, A. Le Tertre, T. Piekarksi, A. Ponce de Leon, A. Ponka, A. Rossi, M. Saez, and J.P. Schouten (1996) 'Methodological Issues in Studies of Air Pollution and Daily Counts of Deaths or Hospital Admissions', *Journal of Epidemiology and Community Health*, vol. 50(1), S3–S11.

U.S. Environmental Protection Agency (1997) *The Benefits and Costs of the Clean Air Act, 1970 to 1990*. Report to the U.S. Congress (Washington, DC: Government Printing Office).

U.S. Environmental Protection Agency (1999) *The Benefits and Costs of the Clean Air Act, 1990 to 2010*. Report to the U.S. Congress (Washington, DC: Government Printing Office).

Venners, S.A., B. Wang, Z. Xu, Y. Schlatter, L. Wang, X. Xu (2003) 'Particulate Matter, Sulfur Dioxide, and Daily Mortality in Chongqing, China', *Environmental Health Perspectives*, vol. 111(4), 562–67.

Viscusi, W.K., W. Magat, and J. Huber (1991) 'Pricing Environmental Health Risks: A Survey Assessment of Risk-Risk and Risk-Dollar Tradeoffs for Chronic Bronchitis', *Journal of Environmental Economics and Management*, vol. 21, 32–51.

Wan, Y. (2005) *Integrated Assessment of China's Air Pollution-Induced Health Effects and Their Impacts on National Economy*, Ph.D. Thesis (Tokyo: Department of Social Engineering, Graduate School of Decision Science and Technology, Tokyo Institute of Technology).

Wang, H., G. Lin, and X. Pan (2003) 'Association between Total Suspended Particles and Cardiovascular Disease Mortality in Shenyang', *Journal of Environment and Health*, vol. 20(1), 13–15.

Wang, H. and J. Mullahy (2006) 'Willingness to Pay for Reducing Fatal Risk by Improving Air Quality: A Contingent Valuation Study in Chongqing, China', *Science of the Total Environment,* vol. 367, 50–57.

Xu, X., J. Gao, D. Dockery, and Y. Chen (1994) 'Air Pollution and Daily Mortality in Residential Areas of Beijing, China', *Archives of Environmental Health,* vol. 49(4), 216–22.

Xu, Z., Y. Liu, D. Yu, B. Chen, X. Xu, L. Jin, G. Yu, S. Zhang, and J. Zhang (1996) 'Effects of Air Pollution on Mortality in Shenyang City', *Chinese Journal of Public Health,* vol. 15(1), 61–64.

Xu, Z., D. Yu, X. Jian, and X. Xu (2000) 'Air Pollution and Daily Mortality in Shenyang, China', *Archives of Environmental Health,* vol. 55(2), 115–20.

Zhang, X. (2002) *Valuing Mortality Risk Reductions Using the Contingent Valuation Method: Evidence from a Survey of Beijing Residents in 1999* (Beijing: Centre for Environment and Development, Chinese Academy of Social Sciences).

2
Why Climate Change Impacts on Agriculture Could be Economically Substantial

Michael J. Roberts and Wolfram Schlenker

For most of human history, agriculture accounted for the dominant share of GDP and employed most labor. Johnson (1997) estimates that in 1800 about 75–80 percent of the labor force in developed nations were engaged in farming, and only 11 percent of the population lived in urban settings (cities with more than 5000 inhabitants). For some of the world, the industrial revolution changed everything. During the 19th century, labor productivity in agriculture (and everything else) increased sharply. By 1980 a unit of labor produced 50–100 times as much wheat or corn as compared to 1800. Productivity growth initially came from machinery replacing human and animal work effort. Since 1930, productivity gains came mostly from development of high-yielding crop species and adoption of intensive farming practices, including use of commercial fertilizers and pesticides. Crop yields (output per unit of land area) increased roughly threefold in the second half of the 19th century, both in the developed and in the developing world. This "Green Revolution" has been attributed more to the efforts of a single man, Norman Borlaug, than to the entrepreneurial efforts of all the world's farmers.

As a result of these changes, many economists have come to believe natural resource scarcity is no longer a constraint on human prosperity. The argument is as follows: productivity gains have far outstripped population growth, causing commodity prices for almost all kinds of agricultural, mineral, and forest products to decline markedly, so that today we consume far more of these naturally derived goods than ever before while they comprise an ever-shrinking share of our economic budget. Prosperity today is not constrained by natural resources, but by physical and human capital, and by government institutions that imperfectly define and protect individual property rights. With population growth slowing, the future looks even brighter. In developed countries, Malthusian predictions died long ago. Even in the poorest countries, most researchers point toward failed institutions as misery's culprit, not toward natural resource scarcity.

Starting from this viewpoint (Johnson 1997, 2000), it is difficult for some economists to surmise how global climate change might lead to large economic consequences. The costs of climate change, according to some, will accrue mainly to the less-developed world. Some authors see the most ominous climate threats stemming from a possible 20-foot sea-level rise that could sink nearly all of Bangladesh, the Nile River Valley, and many small island nations (Arrow 2007). This would both create catastrophic (though transitory) damages and cause massive migration. But many expect climate change impacts on the aggregate costs of producing most goods and services to be small. As Thomas Schelling described the issue in his American Economic Association presidential address:

> Today very little of our gross domestic product is produced outdoors, susceptible to climate. Agriculture and forestry account for less than 3 percent of GDP, and little else is much affected. [...] Manufacturing rarely depends on climate, and where temperature and humidity used to make a difference, air conditioning has intervened. [...] Finance is little affected by climate; similarly for health care, or education, or broadcasting. Transportation can be affected, but improvements in all-weather landing and take-off in the last 30 years are greater than any difference that climate makes. [...] Construction is affected, mainly by cold, and if the average effect is in the direction of warming, construction may benefit slightly. [...] It is mainly agriculture that is affected. But even if agricultural productivity declined by a third over the next half century, the per-capita GNP we might have achieved by 2050 we would achieve only in 2051.

Are the climate change threats to US and world agriculture really so small? We believe it is possible that Schelling and others may be prematurely dismissive about the potential role of climate change on US and world agriculture. Although agriculture currently comprises a small share of both GDP and consumption in the United States and the rest of the developed world, these shares are much larger in developing nations. In 1992, the fraction of consumer expenditures allocated to food still ranged from 50 percent to 67 percent in India, Philippines,

Sudan, and Sierra Leone (Johnson 1997). Moreover, food is an essential good and its demand is highly inelastic. As we will argue below, small shifts in supply or unexpected demand growth can exert large influences on prices, and by simple extension, the share of GDP. Besides, GDP is no welfare measure. As Adam Smith pointed out long ago using his "paradox of value," the price of water is lower than the one of diamonds, even though the former is essential for life while the latter is not. A low price of water is due to its plentiful supply and it does not imply low consumers' surplus.

Recent events serve to illustrate the point as it relates to agriculture: prices for corn, soybeans, wheat, and rice, arguably the world's most important agricultural commodities that supply the caloric basis of our food, all more than doubled between 2005 and 2008. Markets did not perceive these price spikes as temporary phenomena: prices on futures contracts 3 years ahead increased by almost as much as spot prices. Prime agricultural land in the United States doubled in value between 2006 and 2007. These price increases have been largely attributed to US subsidy-induced growth in the demand for ethanol, demand growth for meat in China and India, as well as adverse weather shocks in a few countries, especially the drought in Australia. Still, the quantities at play in these shocks are small relative to world production, so the large price effects speak to the inelasticity of demand, and serve to illustrate the extremely high consumers' surplus associated with agricultural output. In poorer parts of the world, high prices are already having a powerful effect on what and how much people are eating.

Another important consideration is the distribution of worldwide wealth and how it is likely to change over the next century. While demand for agricultural commodities is understandably inelastic, it also shifts out sharply with income growth. The income effect is tied mainly to meat demand. As people become richer they substitute meat consumption for rice, beans, grains, and tubers. Each calorie of meat consumed translates to 5 to 10 calories of required feed grains (mostly corn and soybeans). So basic natural resource demand will grow more rapidly as some of the world's poorer people become richer, even if world population grows more slowly. With populous China and India growing rapidly, per-capita meat demand is growing fast, and with it, demand for feed grains. It is unclear whether a second Green Revolution might serve to keep agricultural production growth on par with aggregate demand growth. If yield growth slows (or does not accelerate), commodity and food prices are likely to rise much further.

The economic implications might vary drastically across the world. In developed countries, only huge price increases are likely to have much influence on food consumption. The price people pay for food in wealthy countries is comprised mostly of labor associated with retailing and transportation and are not much affected by changes in commodity prices (e.g., Leibtag 2008). Even as food prices rise in relation to income, people initially cut back on food consumption away from home (McCracken and Brandt 1987). Substitution of this kind would influence where food is consumed but not how much is consumed. The picture looks different in developing countries: Equilibrium would likely be reached through a decrease in food consumption as poorer people would quickly reach their budget constraint.

It is the combination of a highly inelastic demand for food with large wealth discrepancies between countries that could cause misery and famine in some parts of the world. Climate-induced shocks to agricultural output have the potential to significantly drive up prices which would impact food consumption in poorer areas. How large might these impacts be?

The major agricultural production regions can predominantly be found in temperate climates. As a first proxy, one should hence expect regions that are already warm today to suffer from continued warming while colder regions in higher latitudes might benefit (Rosenzweig and Hillel 1998). So while the impacts on temperate zones are still debated, there is larger consensus that warming will likely be bad for hotter areas. Since developing countries are predominantly found in hotter areas and agriculture constitutes a larger share of GDP in these countries, one may be tempted to focus primarily on developing countries. But this view overlooks the essential fact that world commodity markets are global. Somewhat paradoxically, effects on US agriculture may be the most important for world commodity prices because by almost any measure US agricultural production and exports are the world's largest: the United States produces about 23 percent of the world's calories in the four basic commodities: corn, rice, soybeans, and wheat.

The recent commodity price boom sheds some light on the global importance of US agriculture. Many have attributed the boom to US ethanol policies. These policies simultaneously tax foreign imports of ethanol and subsidize domestic production. This has diverted a significant share of US crops toward the production of corn-based ethanol, which has reduced exports of corn, soybeans, and cotton, thereby creating or at least heavily contributing to commodity price increases. Ethanol policies also seem to have strengthened the link between energy prices and agricultural prices. High commodity prices, in turn, have reverberated throughout the world, markedly increasing the cost of basic food in places like Nigeria (*New York Times* March 9, 2008b), and the FAO reported food riots in Guinea, Mauritania, Mexico, Morocco, Senegal, Uzbekistan, and Yemen (*New York Times* January 19, 2008c). Similarly, large climate impacts on US agriculture would likely to have substantial implications for commodity prices worldwide (*New York Times* April 17, 2008a).

In summary: Those most at risk from large climate impacts are most likely the poor who live in developing nations. But the largest potential impacts are not just those related to changes in climatic conditions *within* these countries, but also repercussions from changes in climatic conditions in *developed* countries because agricultural markets are globally integrated. Part of the calculus includes potential climate change impacts on agriculture worldwide, and how those impacts will ultimately influence commodity prices. Moreover, implications for richer nations could be larger than historical agricultural shares of GDP suggest, since modest production impacts could induce large commodity price changes.

The remainder of this chapter is organized as follows. Section 1 gives a brief history of agricultural production, including historic trends in supply (area versus yield increases) and highlights the important role of the United States as the world's largest producer of the staple food crops. Section 2 focuses on two particular features of food demand: a substitution toward meat with growing income and the low price elasticity of demand, which implies that small quantity changes can lead to big price changes. Section 3 examines the potential impacts of changing climatic conditions on agricultural output in the United States as well as implications for developing countries before Section 4 concludes.

1 Agricultural production, past and present

About three-quarters of the food energy consumed worldwide is derived from just four commodities: rice, wheat, corn, and soybeans. (Cassman 1999 attributes two-thirds of the energy in all human diets to wheat, rice, and corn. Adding the calories for soybean production implies these four crops account for approximately three-quarters of the energy in human diets.) Total caloric production of these four crops from 1961 to 2006 is shown in Figure 2.1. Corn accounts for most calories followed by wheat, rice, and soybeans. Rice is a staple throughout Asia and wheat is used predominantly for bread. Corn and soybeans are used primarily

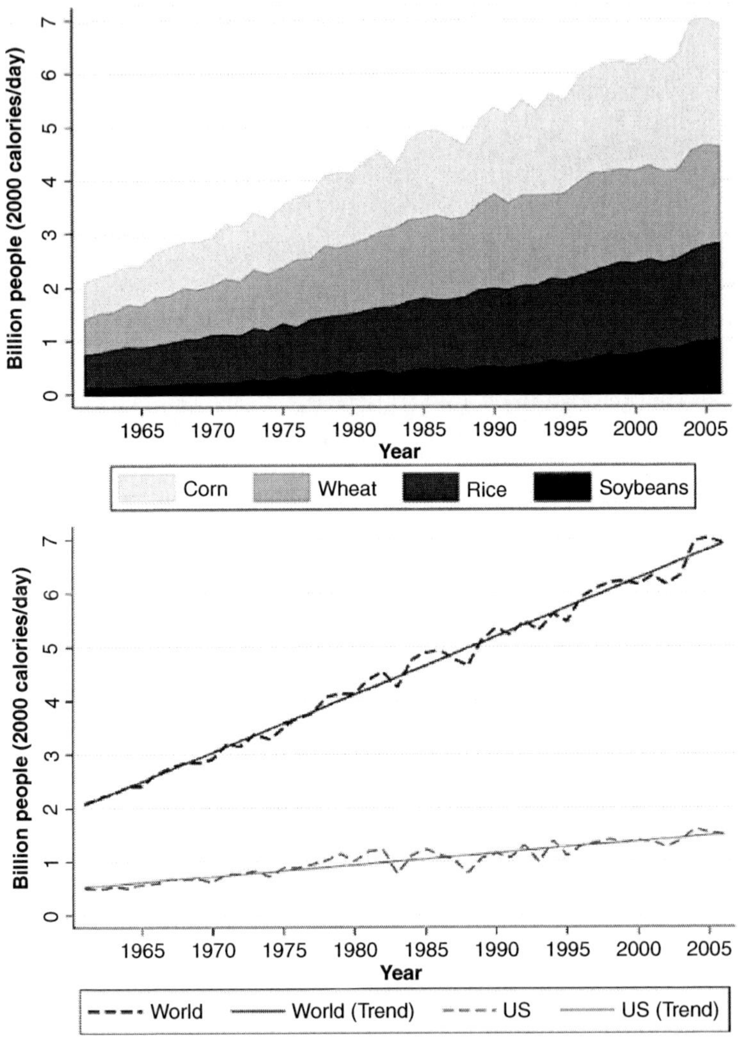

Figure 2.1   Caloric production of the world's largest four crops

Notes: The top plot displays the number of people that could be fed with the caloric production of the world's four largest crops assuming a 2000 calories per day diet. The bottom panel shows world production (black lines) as well as production by the United States (grey lines). The actual time series is shown as a dashed line, while a linear time trend is added as a solid line.

Source: The time series of total production (tonnes) is taken from the Food and Agriculture Organization (2008). The fraction of agricultural yields that is edible as well as the calories per pound edible are taken from Williamson and Williamson (1942).

as feed grains and thus the implicit source of most meat, eggs, and dairy products. Corn meal is also a staple in lesser developed Central American countries and used in a vast array of processed foods in developed countries. In recent years, a large share of corn produced in the United States has been diverted to the production of ethanol which, besides fuel, yields distiller's grains, which may be fed to ruminant livestock.

1.1 Sources of supply growth: Area versus yields

World caloric production by the five largest producers as well as harvested area are shown in Figure 2.2. While total production shows a strong upward trend over the last 40 years, total acreage has grown by a much lesser amount. This implies that production increases are mainly due to yield increases (output per acre) rather than an expansion of cropped land area. Output growth happened on the intensive margin, not the extensive margin. The exception is soybean production, particularly in Argentina, Brazil, and the United States over the last decade, where acreage has expanded considerably. One reason is that soybeans are used as a rotation crop with corn in the United States. The other reason is that production increases in South America have been matched by imports of soybeans into Asia, particularly China. These imports are used for soybean oil and as feed for livestock, which has been expanded considerably as shown in Section 1 below. The United States is among the five largest producers for three of the world's four largest crops. Figure 2.2 shows that the United States is the largest producers of corn and soybeans with a market share of approximately 40 percent, while it is the third largest producer of rice with a market share of 9 percent.

Figure 2.3 shows average yield as well as area planted for the four major staple crops in the United States from 1866 to 2007. Similar to world agricultural trends, increases in yields outpaced increases in acreage in the second half of the 19th century. For corn and wheat, the harvested area was lower in 2007 than what it was in 1925, yet total production was up due to a dramatic rise in yields. For rice the production area continued to increase, but the growth in yields was faster than the growth in acreage. The exception is, again, soybeans, a crop that has become more popular as feed crop in recent decades and whose area continued to increase throughout the 19th century. Soybean acreage, however, still remains below that of corn and accounts for less calories than corn (see Figure 2.1 above).

However, the story has been very different historically. While yields improved slightly between 1866 and 1925 for corn, rice, and wheat in the United States, production increases were driven largely by expansion

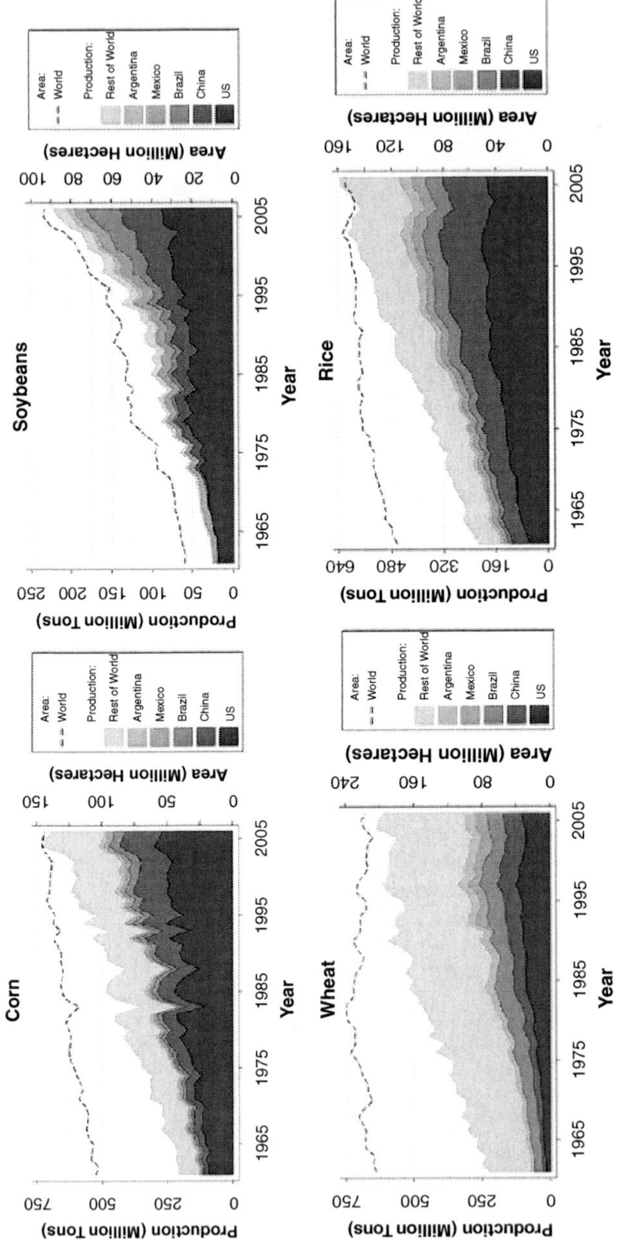

Figure 2.2 Production and area by top five producers

Notes: Graphs display total production of the world's five largest producers from 1960 to 2005 as shaded area, while the top area is the production of all remaining countries. The total area is added as a dashed line.

Source: The time series of total production (tonnes) and acres planted (hectares) is taken from the Food and Agriculture Organization (2008). The fraction of agricultural yields that is edible as well as the calories per pound edible are taken from Williamson and Williamson (1942).

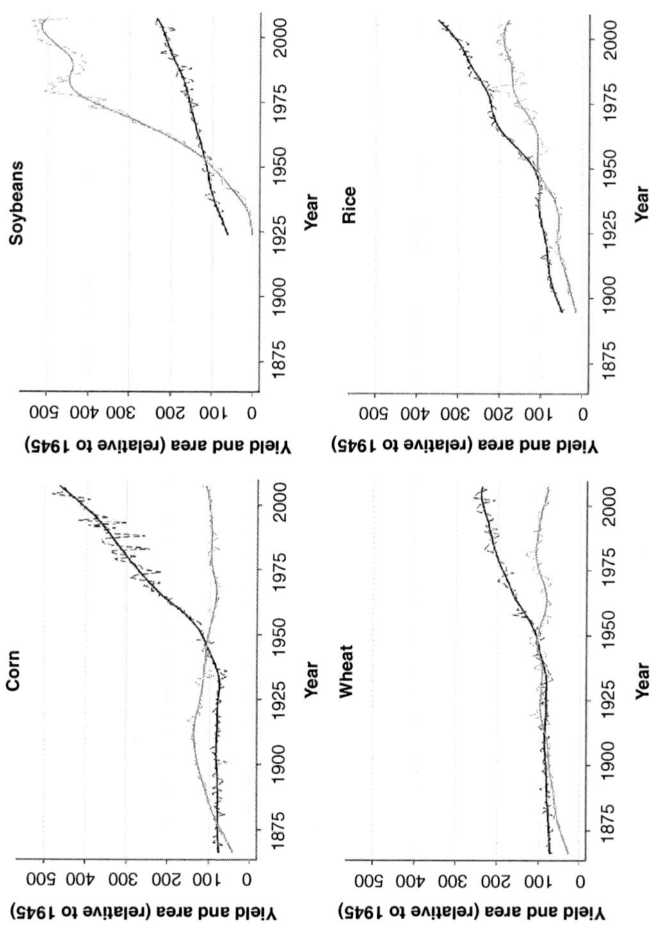

Figure 2.3 Average yield and growing area in the United States over time

Notes: Graphs display time series of average yields (black line) and growing area (grey lines) in the United States from 1866 to 2005. All time series are normalized relative to 1945. Thin dashed lines display the time series, while thick solid lines show smoothed trends (using the Epanechnikov kernel with a bandwidth of one decade).

Source: National Agricultural Statistics Service (2008a).

of land devoted to these crops. Area harvested increased by roughly a factor of three for corn and wheat between 1866 and 1925 and rice between 1895 and 1925. Growth in output occurred mainly on the extensive margin, through an increase in the production area. A similar pattern holds for other countries: Production increases initially were driven primarily by expansion of the acreage. The Green Revolution in the second part of the 19th century resulted in roughly a tripling of output per area and hence most production increases occurred on the intensive margin.

Future supply increases might come both from the intensive margin (new crop varieties, e.g., genetically modified foods) or from the extensive margin (expansion in growing area). Searchinger et al. (2008) argue that diverting corn for ethanol production will drive up prices and lead to a conversion of forests to agricultural land and this land conversion will result in significant carbon emissions.

1.2   Production in the United States

As mentioned above, the United States is the largest producer of agricultural commodities. It produced on average 23 percent of the world's calories in the years 1961–2006 as shown in the top panel of Figure 2.4 even though its share of the world population is only around 5 percent. The larger portion of production is used domestically as the United States consumes more than its population share of the four key commodity crops, mainly because of its relatively high consumption of meat, where animals consume a multiple in calories than what they produce. Despite its high consumption, US predominance in agricultural production makes it, by far, the world's largest agricultural exporter of agricultural commodities. The bottom panel of Figure 2.4 illustrates this predominance for the four key crops. It displays the fraction of domestic production that is exported. The United States exports slightly less than one third of its corn and soybeans, and around half of its wheat and rice production. The United States is also a leading producer and/or exporter of many specialized crops, including cotton, oranges, tobacco, strawberries, almonds, pistachios, wine grapes, and a host of other high-value fruits and vegetables. We do not focus on these high-value crops because they constitute a much smaller fraction of overall caloric input and hence are less critical for global food supply in the long run.

US agricultural predominance stems from its large and rich land resources, temperate climate, advanced production systems, and exceptionally large output. Thus, even while agriculture comprises a small share of GDP in the United States, its production is critical to world

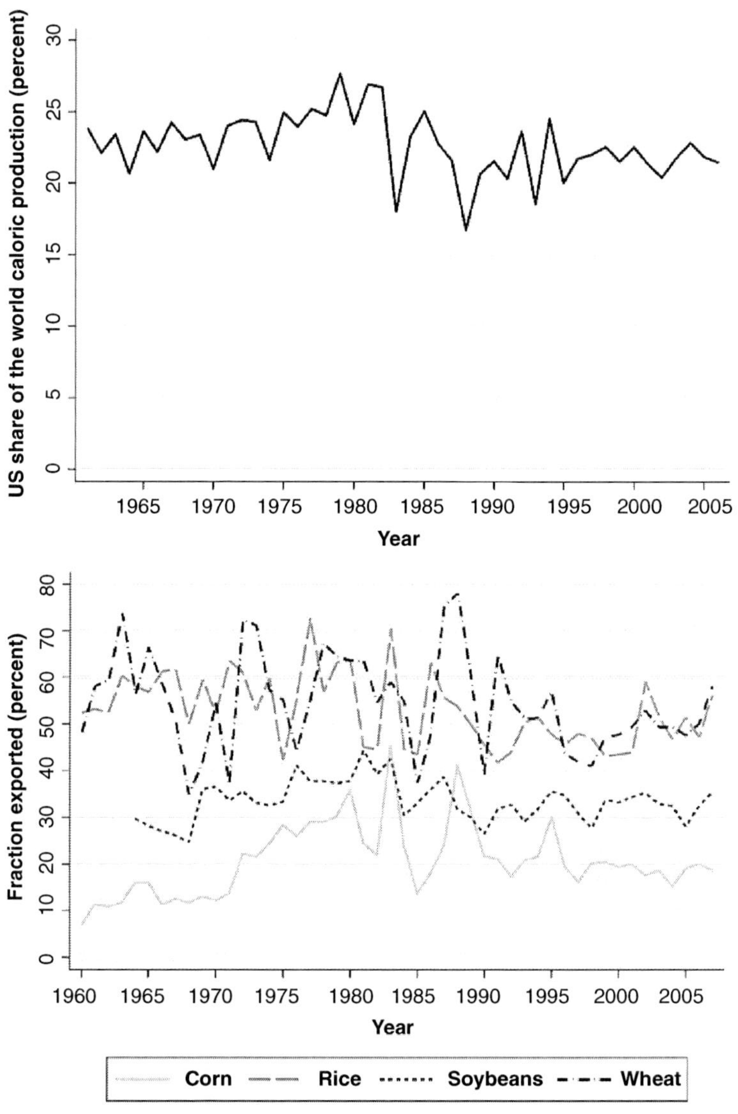

Figure 2.4   US share of world calorie production (corn, rice, soybeans, and wheat)
Notes: The top plot displays the share of global calories in corn, rice, soybeans, and wheat produced by the United States. The bottom panel displays the fraction of US production that is exported.
Source: The time series of total production (tonnes) is taken from the Food and Agriculture Organization (2008). The fraction of agricultural yields that is edible as well as the calories per pound edible are taken from Williamson and Williamson (1942). Data on exports is taken from the Foreign Agricultural Service (2008).

food supply and agricultural prices. Indeed, the United States plays about twice the role in world agricultural production than Saudi Arabia does in world oil production: The United States produces 23 percent of the world's calories from the four basic commodities while Saudi Arabia supplies about 12 percent of the world's oil. The United States is also a leading producer of a wide array of specialized crops like oranges, strawberries, almonds, and pistachios, among many, many others.

One notable feature of the data plotted in Figure 2.1 above is that variability around the upward trend in world caloric production comes predominantly from variations in corn production. A natural explanation for this phenomenon is that corn production is more geographically concentrated than the other three crops. As mentioned above, the United States accounts for about 40 percent of the world's total corn production. Within the United States, corn and soybean production (as well as some wheat production) is concentrated in a confined geographic area. Figure 2.5 displays a plot of the cropland area in the United States from satellite scans. The Midwestern states of Iowa, Illinois, Indiana, Ohio, and, to a somewhat lesser extent, Missouri, Wisconsin, Kansas, and Minnesota, clustered together in the middle of the country, can experience similar weather and yield outcomes, which give rise to some aggregate variability. Since production of the other crops is more spread out across countries and continents, weather-related yield shocks tend to average out, resulting in less aggregate variability.

A consequence of US production being both large and geographically concentrated is that weather-related yield shocks in the United States have powerful effects on worldwide output. This is illustrated in Figure 2.6, which relates worldwide deviations in caloric production from a linear time trend to deviations in US caloric production. The correlation coefficient between the two is 0.80, and the regression coefficient is 1.01 (t-value 8.72), suggesting that there is a one-to-one relations between US deviations and world deviations. Since the United States produces around 23 percent of the world's caloric output, a 10 percent change in caloric production in the United States is predicted to change worldwide caloric production by roughly 2 percent. Because of its size and concentration, production shocks in the United States have implications for world supply. If climate change were to significantly alter supply in the United States, we would expect it to have repercussions for world prices. How much prices are predicted to increase depends not only on supply but also on agricultural demand, which we examine in more detail in the next section.

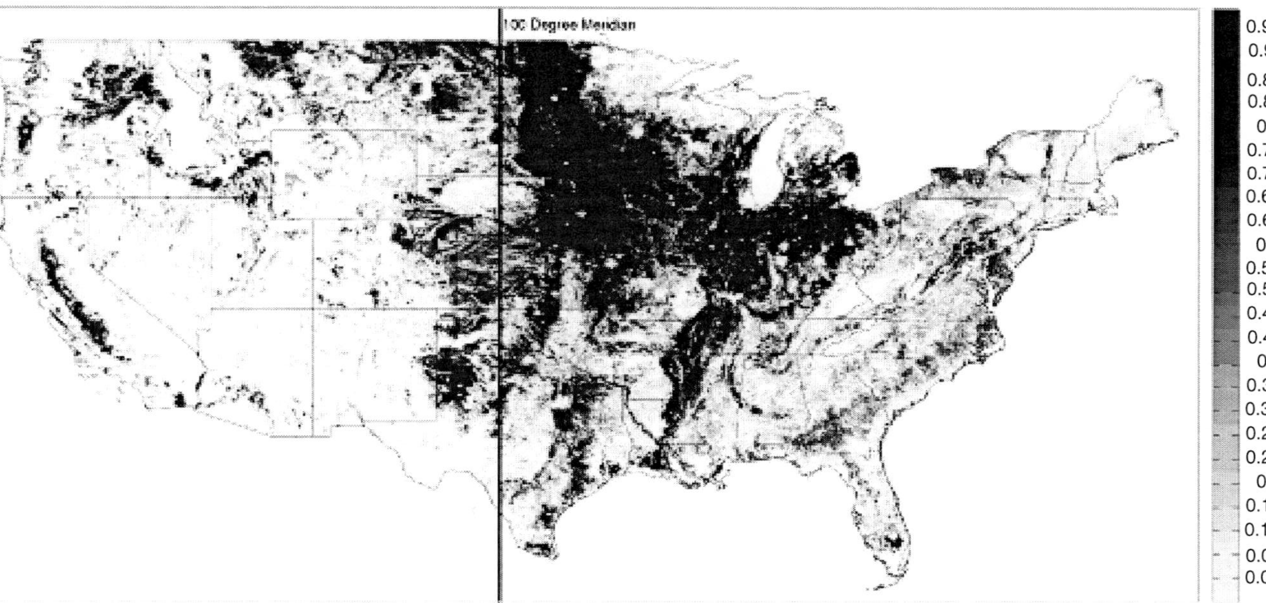

Figure 2.5   Agricultural area in the United States

Notes: Graph displays the agricultural area in the United States. Each pixel indicates the fraction of the 2.5 × 2.5 mile grid that is cropland area. The darker the grid color, the larger the fraction of a county that is cropland area. White areas have no cropland. The 100 degree meridian is 100 degrees west of Greenwich, England.

Source: LandSat satellite scan.

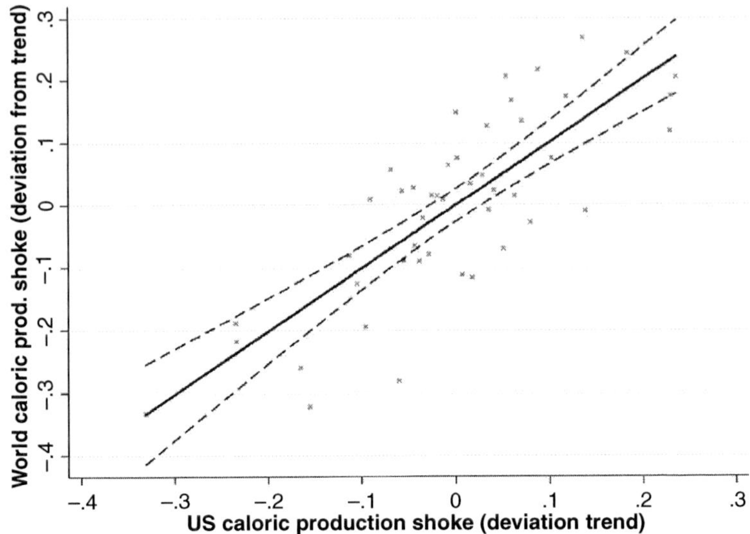

Figure 2.6   Relation between US production shocks and world production shocks
Notes: Graph correlates total deviations of calorie production from a linear time trend in the United States and the entire world as shown in the bottom panel of Figure 2.1. A regression line (solid line) as well as the 95 percent confidence bands (dashed lines) are added. The regression coefficient is 1.01 with a *t*-value of 8.72.
Source: The time series of total production (tonnes) is taken from the Food and Agriculture Organization (2008). The fraction of agricultural yields that is edible as well as the calories per pound edible are taken from Williamson and Williamson (1942).

2   Agricultural demand

Two features of agricultural demand are particularly important for food scarcity in the future: the income elasticity of the demand for meat (which requires a multiple of calories as an input than it produces) as well as the price-elasticity of demand for agricultural goods.

2.1   Meat demand and income

As people grow richer, food demand shifts from an almost exclusively vegetarian diet to one comprised increasingly of meat. The top panel of Figure 2.7 shows a cross-country analysis in 2003: average meat consumption is plotted against the log of GDP in a country, where the size of a circle is proportional to the country's population. Richer countries consume far more meat per capita than poorer countries. This suggests much of the world's population, including China and India, are likely to increase meat consumption further as their incomes grow. This growth

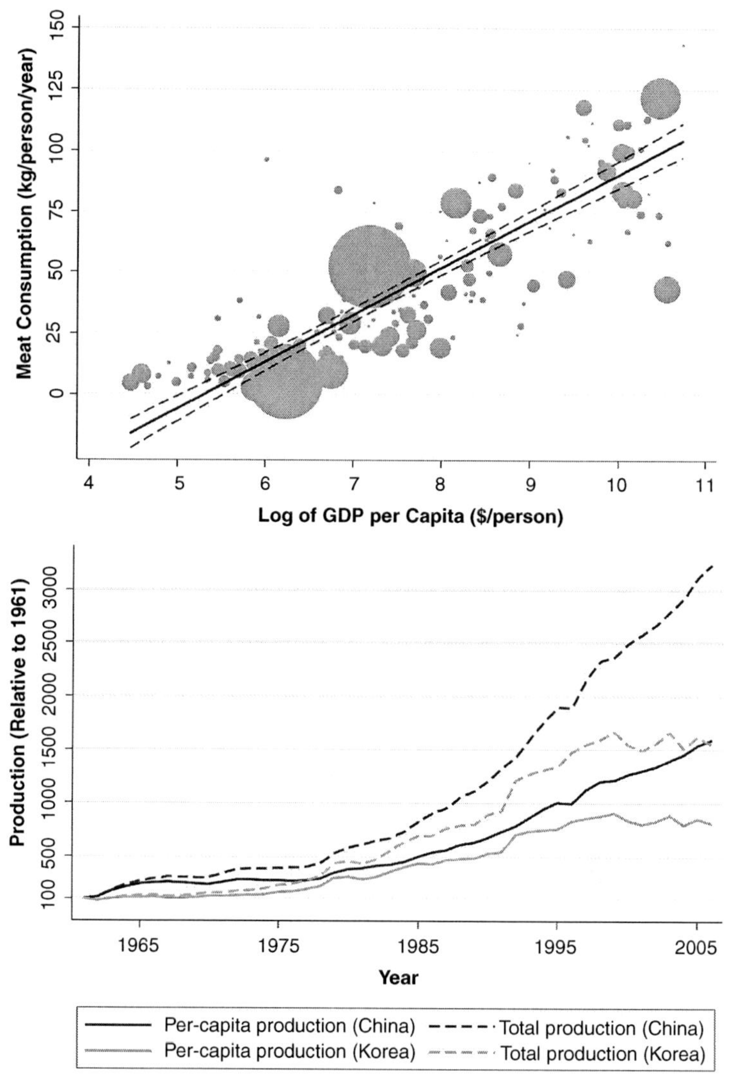

Figure 2.7   Meat consumption and production as a function of income

Notes: The top panel displays average meat consumption across countries in 2001–2003 as a function of log GDP per capita in 2003. The size of a dot is proportional to the population of a country. A regression line (solid line) as well as 95 percent confidence bands (dashed lines) are added. The bottom panel displays the increase in meat production in China (black lines) and South Korea (grey lines). Total production is shown as a dashed line, while per-capita production is shown as solid line. All lines are relative to 1961, the start of the time series.
Source: Meat consumption is taken from the Food and Agriculture Organization (2007). Meat production in China and Korea is taken from the Food and Agriculture Organization (2008).

will be the largest source of commodity demand growth in the coming decades. Meat requires far more caloric input than a vegetarian diet. Animals consume feed crops (often corn and soybeans) while they are raised. Most of the calories animals consume are expended before the animals are slaughtered. Gerbens-Leenes et al. (2002) estimate that it takes 3–5 times the land area for meat production than for a vegetarian diet in the Netherlands. Others have estimated that it takes 5 to 10 calories of feed crops to produce a single calorie of meat. Continued growth in meat demand will only increase demand for staple food crops.

A comparable relationship between meat demand and income holds within countries over time. Worldwide meat production increased from 71 million tons in 1961 to 273 million tons in 2006. In China alone meat production grew from 2.5 million tons to 82 million tons in this time period (Food and Agriculture Organization 2008), which is mainly consumed domestically. The large growth in production stems from its massive population and rapidly growing income. Population growth, however, is just a small part of the story. While total population doubled over this period in China, total meat production increased 33-fold as shown in the bottom panel of Figure 2.7. As a result, China produced and consumed roughly 30 percent of the world's meat in 2006. South Korea's meat production has increased over the same time period but leveled off during the 1990s. In 2006, per-capita meat production (tons of meat) in Korea was slight more than half the number in China.

It is important to note that if increasing demand for meat is not matched by increased yield growth and commodity supply, growing scarcity will drive up commodity and land prices. While higher prices normally act to reduce the quantity demanded via the demand curve, we will show below that demand is extremely inelastic. Thus, equilibrium would be reached with commodity and meat prices much higher than they are now. In richer countries, higher prices may have a positive side effect of inducing healthier diets. With more modest meat consumption, there should be plenty of food to feed the world's population and there is no reason to expect that higher food prices would lead to meaningful limits to continued economic growth. But continued growth in meat consumption could increase the scarcity of staple crops used as feed.

2.2   Price-elasticity of demand for food

While income-growth shifts out the demand for meat and meat consumption requires more calories from staple crops than consuming them directly, the price effect of the increased demand for staple crops crucially depends on the elasticity of demand. Since food is essential for life, one would expect the demand to be inelastic. In wealthier nations demand for food commodities is inelastic because they account for a small share of income. Commodity price and quantity fluctuations bear out the inelasticity of demand empirically.

The top panel of Figure 2.8 plots real prices for the four key commodities from 1961 to 2008 as well as the calorie-weighted average. Prices for 2008 are preliminary, reflecting futures prices from April 1, the time this chapter was written. The average price uses the average production share of each crop in Figure 2.1 over the period 2000–2006. All prices are in caloric units to match Figure 2.1 above. The vertical axis gives the price of annual caloric need for one person (assuming 2000 calories per day) if all calories were derived from consuming the raw commodity. While the overall trend has been downward, commodity prices fluctuations are large, and they tend to move up and down together. The large price fluctuations bear a striking contrast to the relatively small quantity fluctuations plotted in the bottom graph of Figure 2.1 above. Thus, it would seem that relatively small quantity changes lead to big price changes. The price fluctuations appear especially large given all four of these commodities are highly storable, which means accumulation and depletion of inventories smoothes consumption relative to production, and thus should smooth prices as well. Because of storage, price changes also tend to be very persistent, with near unit roots. Thus, the large price fluctuations would seem to indicate that demand and/or supply are extremely inelastic.

Since US fluctuations in production explain most worldwide variations, it seems reasonable to infer that much of the short-term fluctuations in quantities are due to weather. Thus, we might reasonably assume that deviations from trend aggregate production are transitory and unexpected shifts in supply. Following this assumption, we can obtain a crude estimate for the world demand elasticity by regressing deviations from the trend in world caloric production (plotted in the bottom panel of Figure 2.1 above) against changes in the average caloric price. This regression and its associated scatter plot are given in the bottom panel of Figure 2.8. Because prices are on the left-hand-side of the regression and presumably random quantity fluctuations are on the right, the relationship provides insight into the inverse demand. The statistically significant slope is -3.17 (with a t-value of 3.69) and the implied demand elasticity is the inverse, or -0.32.

Interpreting this regression line as a demand curve assumes all deviations from the quantity trend are unanticipated supply shocks. If some

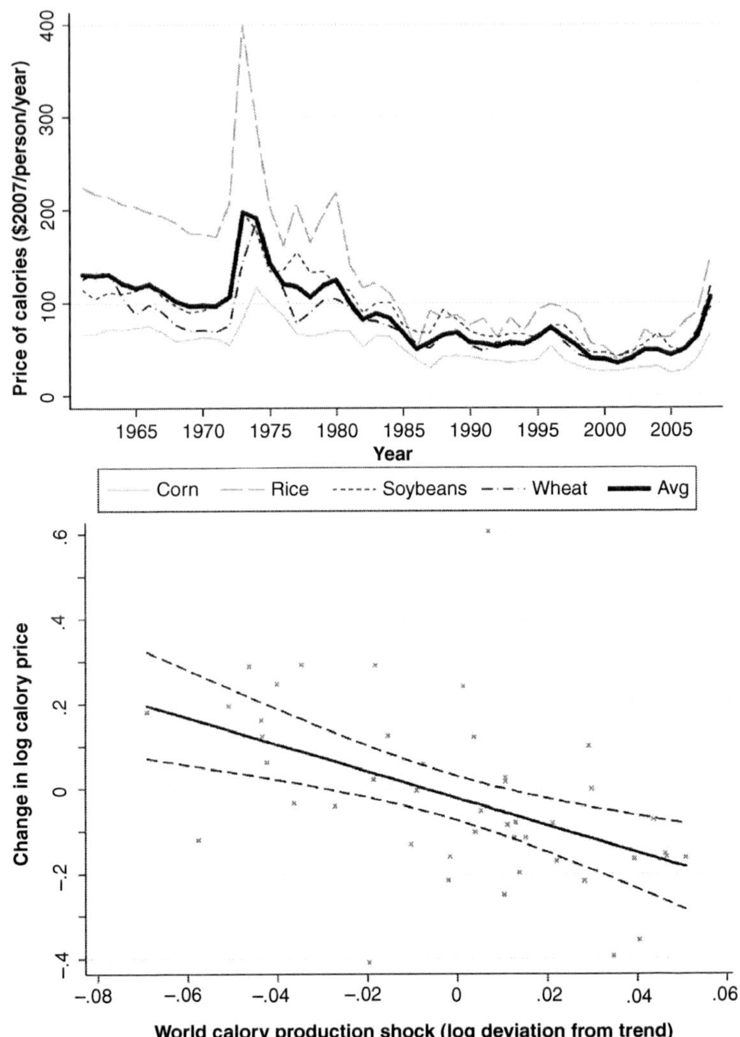

Figure 2.8   Prices and inverse demand function

Notes: The top panel displays real prices (US$ 2007) of a 2000 calories/day diet derived from the four major crops for the years 1961–2008. The bottom graph displays changes in average log commodity prices as a function of production shocks (log deviations in world caloric production from a linear time trend). The regression line is added as a black line, and the 95 percent confidence band is added as dashed lines.

Source: Price is taken from the National Agricultural Statistics Service (2008b). The time series of total production (tonnes) is taken from the Food and Agriculture Organization (2008). The fraction of agricultural yield that is edible as well as the calories per pound edible are taken from Williamson and Williamson (1942).

of the quantity deviations actually stem from demand shocks, or if some of the deviations are anticipated (e.g., observed area expansions at the time of planting), then the demand elasticity is biased upward in magnitude. That is, the price response is too small relative to the true inverse demand curve. More importantly, because most supply shocks are buffered by accumulation or depletion of inventories, only a small fraction of a transitory quantity shock will be accrued in the current period. In Fisher et al. (2007) we show that nearly 100 percent of a typical corn yield shock is buffered by accumulation or depletion of inventories in the United States. As a result, the elasticity is biased upward in magnitude if one does not account for storage, and hence demand is less elastic than in our exercise. The inescapable conclusion is that the underlying demand elasticity for basic commodity calories is extremely inelastic. In Roberts and Schlenker (2009) we use the world-wide sum of local yield shocks as a proxy for weather-induced supply shocks, account for storage, and estimate a demand elasticity of -0.05. Thus one can expect a doubling of prices from a 5 percent downward shift in supply.

Another way to approximate the demand elasticity is to consider the recent rise in prices. Around the year 2000, basic annual caloric need could be bought for 39 dollars (production weighted average price in Figure 2.8). At this writing, in April of 2008, the price of basic annual caloric need has risen to $150 for rice, $117 for wheat, $96 for soybeans, and $67 for corn in 2008. The production-weighted average price has risen to $106, more than twice the level in 2000. Much of this increase has been attributed to ethanol subsidies. These subsidies, together with high oil prices, have grown corn-based ethanol production from almost nothing in 2000 to a projected 9.3 billion gallons in 2008/2009. Current ethanol production accounts for about 30 percent this year's US corn crop and is projected to account for about one-third of the US crop over the next decade. Given US corn accounts for about 40 percent of corn produced worldwide, and corn accounts for roughly one-third of calories produced worldwide (Figure 2.1) in the four essential crops, corn-based ethanol production will have soon displaced about 4.4 percent of the calories that would have been directed toward the food supply. Thus, the predicted increase in prices would be $4.4/0.032 = 139$ percent, which is nearly as high as the 175 percent price increase we observed for the production-weighted average price of a calorie between 2000 and 2008.

We note that these simple calculations do not account for supply response induced by higher prices (i.e., movement along the supply curve). Less than perfectly inelastic supply would dampen the price effect of growing ethanol demand. In Roberts and Schlenker (2009) we argue that supply is more elastic than previous empirical estimates.

2.3   Implications for poor countries

Market mechanisms ensure that a scarce resources end up in the hands of those most willing and able to pay for them, thereby ensuring Pareto efficiency. Yet, there could be considerable distributional concerns if some grow richer while others remain poor. Today about half the world lives on less than \$2/day or about \$670/year. At the same time, many in wealthy countries are willing and easily able to pay \$670/year, perhaps several times over, to satisfy preferences for a diet rich in meat. If some grow much wealthier, consume more meat and commodity resources, and thereby drive up commodity prices, we may find that those remaining poor can no longer afford basic caloric requirements. Indeed, at today's rice prices it requires \$150/year, a fourth or more of half the world's budget, just to satisfy basic caloric needs, not including retailer markups, tariffs, or transportation costs. At this writing, the *New York Times* and *Washington Post* are reporting that the high price of food staples is causing food riots in Yemen and Morocco and hoarding of rice in Hong Kong, while governments of many less developed nations are banning exports. While the current situation appears to stem in part from the recent ethanol boom, it is easy to see how high prices might be sustained or rise further from growing meat demand.

The possibility of high food prices leading to widespread famine described here may seem Malthusian in nature. It is not. Malthus argued that population growth would always exhaust economic growth in the long run, leading most to live in misery on the brink of survival. Rather, we believe it likely that if widespread famine were to occur it would happen simultaneously with continued per-capita income growth. Prosperity would simply accrue to the richer half or two-thirds of the world while poorest third struggle to survive. Moreover, it is not a misery driven by population growth as much as growing income inequality coupled with less-than-spectacular technological progress in agricultural production.

3   Potential climate impacts

In the past two sections we have given a brief overview of agricultural markets. Our goal was to provide some context for potential climate impacts by characterizing global agricultural markets, how they have changed over time, and how they might be expected to change in the future. While agriculture remains a tiny share of economies in developed nations, that tiny share, particularly in the United States, has large potential welfare implications for those in developing nations. Specifically, the growing demand for meat is making basic staple foods (which are used as feed stock in meat production) scarcer, as is the demand to use these staple crops for biofuel production. Even though only a small share of caloric production was shifted toward ethanol and meat in the last few years, prices have more than doubled for these commodities. If climate-induced yield declines would add further scarcity by reducing output, prices might skyrocket even further with large consequences for poorer countries where food constitutes a large share of income even at today's prices.

Most research to date suggests that anticipated climate changes will harm agricultural production in the developing world (Lobell et al. 2008). The poorest countries also tend to be among the warmest countries, where soils and climates are already suboptimal for production of agricultural staples. Further warming is likely to hurt productivity in these countries. Research on the United States and similarly temperate climates is mixed. Previous studies have found a wide range of possible climate-change impacts on agriculture, ranging from significant benefits to large damages (Mendelsohn et al. 1994, Darwin 1999, Schlenker et al. 2006, Kelly et al. 2005, Timmins 2006, Ashenfelter and Storchmann 2006, Deschênes and Greenstone 2007). Given the large US role in world commodity markets, it is important to focus research efforts on the United States in order to reconcile these mixed findings.

Previous statistical studies differ on two key dimensions: (i) the statistical identification strategy used in the analysis (e.g., cross-sectional variations in climate versus year-to-year fluctuations in weather), and (ii) different functional relationships between temperature and precipitation variables and agricultural output or farmland values. It is our view that differences in previous results are primarily due to the latter, *not* whether one uses the cross-section or the time series as the source of identification. If the weather variable is modeled adequately, both a panel of yields and a cross-sectional analysis of farmland values give comparable results.

In Schlenker and Roberts (2009) we show that the relationship between yields and temperatures is highly nonlinear for corn, soybeans, and cotton. The study develops a new fine-scaled weather data set that incorporates how much time a crop is exposed to each 1°C interval during each day of the growing season to precisely estimate this nonlinearity. Yields are increasing in temperature until about 29°C for corn, 30°C for soybeans, and 32°C for cotton, but become very harmful above these thresholds. The slope of the decline above the optimum is significantly steeper than the incline below it. While warming below the threshold is hence desirable, it is very harmful if temperatures exceed these thresholds. Given the difference in slopes, it is predominantly the increase in very hot days that drives most of our impact estimates. The predicted impacts at the end of the century are large and highly significant ranging from a 39 percent to a 53 percent decline in yields for the three crops in question under the B2-emission scenario in the Hadley III model. The B2-emission scenario is a mid-range emission scenario.

We obtain comparable results when we look at the pure time-series of aggregate yields or the cross-section of average yields. The latter incorporates farmers' responses to differences in long-run weather averages (climates) as farmers in the south can expect that temperatures will be warmer than in the north. The former relies on year-to-year weather fluctuations that are unknown at the point of planting and hence are difficult to adapt to. The fact that both give similar results suggests that adaptation possibilities *within a crop species* are limited.

Another approach to measure adaptation possibilities (e.g., crop switching) is to estimate a reduced-form relationship between farmland values and average climate in a county. In an efficient market, farmland values will equal the discounted profits a farmer can obtain from a piece of land if it is put to its best use. If it is more profitable to grow oranges in a warmer climate than to grow corn, farmland values will equal the discounted profit from planting oranges. Hence, if two plots of land of identical quality have different values in different climates, the differences can be attributed to the difference in climate. Such a hedonic regression is a partial-equilibrium analysis and hence implicitly assumes that prices remain constant. Following our earlier discussion this is a strong assumption, and the results should hence be seen as the capitalized value of production changes under current prices. The highly inelastic demand for food implies that farmers might actually benefit if reductions in output are more than offset by price increases.

The empirical challenge of the hedonic approach is to find two plots of land that are identical except for differences in climates, that is, the researcher has to account for other confounding factors. If an omitted variable that influences farmland values is correlated with climate, the coefficient on the latter will be biased as it picks up the effects of the omitted variable. We will show below that the results are fairly stable under a wide set of sensitivity checks. While there is no test to show that omitted variables are not a problem, the fact that the results are insensitive to set of controls is at least partially reassuring. If an omitted variable is correlated with one of the controls that is included in the regression, the control variable will pick up some of the effect that is attributable to the omitted variable. If coefficient estimates are stable when controls are included and excluded, the omitted variable would have to be correlated with climate but not one of the controls. Hence, showing that the results are insensitive if one includes and excludes certain controls (e.g., soil quality) reduces the set of possible omitted variables that might bias the results, as these would now have to be correlated with climate but uncorrelated with the control that was included.

To check the sensitivity we estimate the cross-sectional relationship between average farmland values in a county and climate for each of the eight census years 1969, 1974, 1978, 1982, 1987, 1992, 1997, and 2002 by including/excluding each of the following eight control variables: (i) latitude, (ii) income per capita, (iii) population density, (iv) average water capacity of the soil, (v) k-factor of the top soil layer, (vi) minimum permeability of all soil layers, (vii) percent clay, and (viii) the fraction that is classified as top soil. These are the controls used in Schlenker et al. (2006). We also include/exclude the square of each of these eight control variables and run the regression with and without state fixed effects for a total of $8 \times 2^8 \times 2^8 \times 2 = 1{,}048{,}576$ permutations. Each permutation includes the same five climatic variables: three degree days variables (explained below) and two precipitation variables, but varies the set of control variables and the year in which the relationship is estimated. The kernel density of predicted climate change impacts is shown in Figure 2.9 and the density of the regression coefficients on the five climatic variables is shown in Figure 2.10. Degree days are truncated temperature variables to capture the nonlinear relationship between temperature and agricultural output. For example, degree days 8–32°C are the daily degrees above 8°C, but below 32°C, that is, a temperature of 10°C would be 2 degree days, and any temperature equal to or above 32°C would be 24 degree days. Similarly, degree days above 34°C are temperatures in excess of 34°C. In each of these two figures the top row displays the results if we use the bounds of the agronomic literature that were used by Schlenker et al. (2006) with a degree days measure that was derived using monthly data. Thom's rule is an interpolation routine that relates the standard deviation between months to the daily variance in temperatures. In a follow-up study we replace Thom's interpolation method of monthly temperatures with daily values (Schlenker and Roberts 2006). The second row of Figure 2.9 and 2.10 replicate the analysis with daily temperature values but continues to use the same bounds. Finally, the last row uses daily values with the optimal bounds for corn from Schlenker and Roberts (2009).

Under all specifications and data sets predicted climate change impacts by the end of the century under the B2-scenario are unambiguously negative. The 95 percent confidence interval of our roughly 1 million regressions spans from approximately 40 percent to 80 percent among the three graphs in Figure 2.9. This effect is driven by the predicted increase in extremely warm temperatures. The coefficient on the degree days above the threshold is unambiguously negative as shown in the left column of Figure 2.10. By the same token, the coefficient on the two precipitation variables (last two columns of Figure 2.10) does not switch signs for the intermediate 95 percent of the values of our sensitivity checks. The sign on the degree days variable capturing intermediate temperatures (8–32°C and 10–29°C) appears less robust to the inclusion or exclusion of control variables, especially once we use daily values in the construction of the variables.

While these production responses are large and significant given that the United States produces 23 percent of the world's calories, a word of caution might be in order. First, our regression is based on past observations and cannot account for CO_2 fertilization or new crop varieties. Moreover, reductions in US production might be offset by production increases in areas further north, especially Russia which has good soils, but some authors argue that vast areas that are now frozen over will become swamps after an increase in temperature that is ill-suited for farming, Whether substitute production is feasible (through new crops or new area) is key in determining whether foods will become more scarce and expensive given the negative effects we have outlined. Potential impacts not only depend on the demand elasticity, but also the supply elasticity. The more elastic the supply, the larger the production response and the lower the price effect. However, the fact that futures prices 3 years from now equal current spot prices suggests that immediate supply responses through enlargement of the agricultural area are not anticipated by the market, despite the recent run-up in prices.

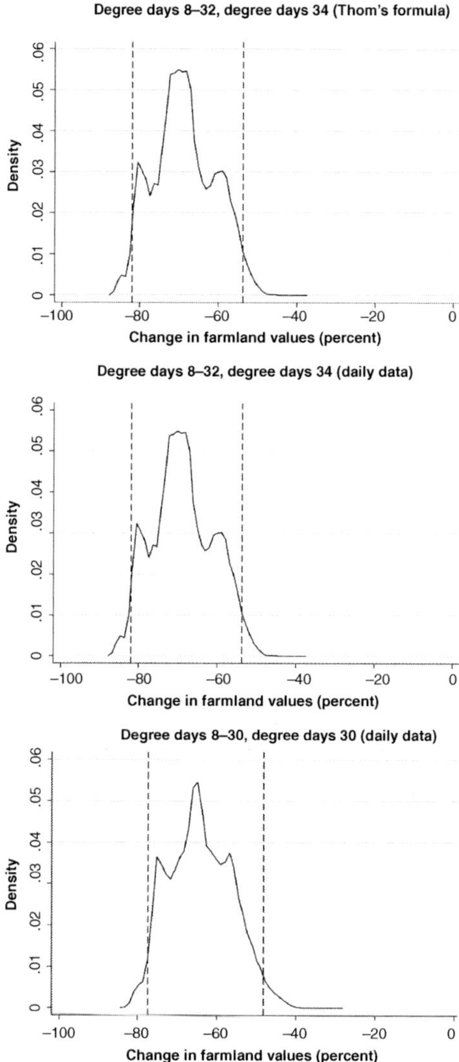

Figure 2.9 Robustness of climate change impacts—Hedonic regression

Notes: Graphs display kernel densities of predicted climate impacts under 1,048,576 model permutations. The 95 percent confidence interval is added as dashed lines. The top graph uses the specification of the climate variables from Schlenker et al. (2006), that is, a quadratic in degree days 8–32°C and the square root of degree days above 34°C as well as monthly data. The middle graph uses the same climatic variables and bounds but uses daily data to derive degree days instead of monthly averages (Schlenker and Roberts (2006)). Finally, the bottom graph uses the optimal bounds obtained in Schlenker and Roberts (2009) for corn together with daily data.

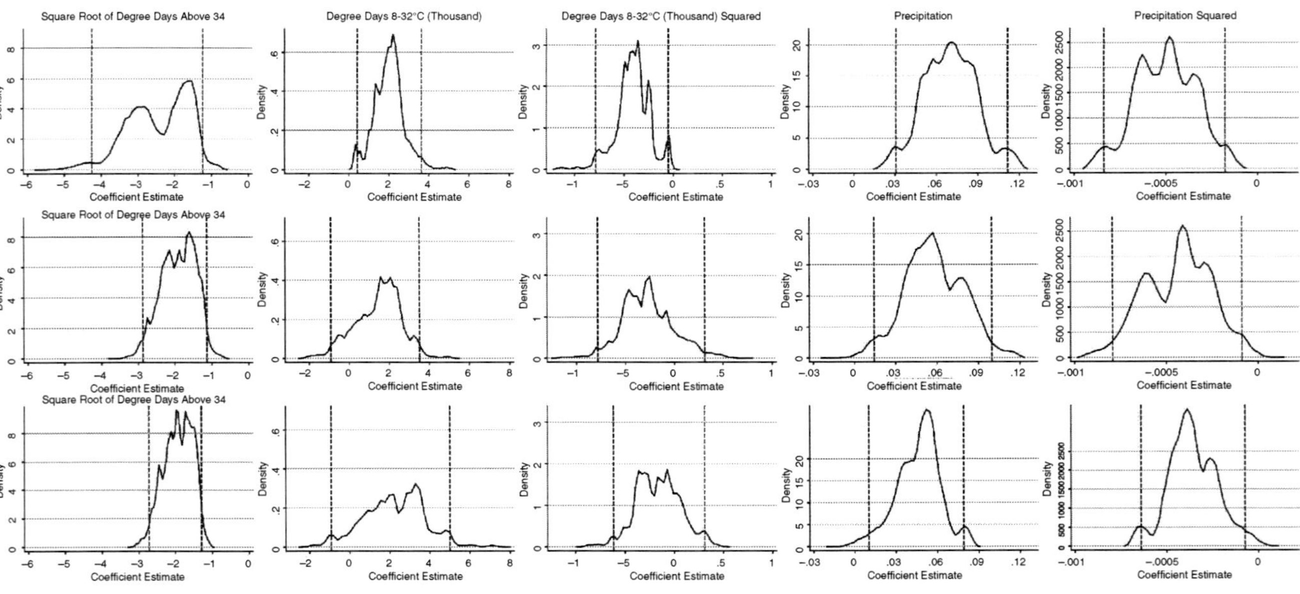

Figure 2.10 Robustness of coefficient estimates—Hedonic regression

Notes: Graphs display kernel densities of regression coefficients under 1,048,576 model permutations. The 95 percent confidence interval is added as dashed lines. The graphs in the top row use the specification of the climate variables from Schlenker et al. (2006), that is, a quadratic in degree days 8–32°C and the square root of degree days above 34°C as well as monthly data. The graphs in the middle row use the same climatic variables and bounds but uses daily data to derive degree days instead of monthly averages (Schlenker and Roberts (2006)). Finally, the graphs in the bottom row use the optimal bounds obtained in Schlenker and Roberts (2009) for corn together with daily data.

4 Conclusions

Research investigating potential climate change impacts on agricultural production has generally found that a warmer climate will harm yields throughout much of the developing world. While findings from research in more temperate climates are mixed, our own research on the United States, the world's largest agricultural producer, indicates potential impacts are both negative and substantial. If our predicted climate-change impacts were to bear out, price and welfare effects would likely be considerable, especially if worldwide income inequality persists into the future. Large income inequality, coupled with highly inelastic demand for food commodities, means that even modest negative climate change impacts could make food unaffordable to the poor. Widespread and persistent famine would seem plausible.

Although the more dismal possibilities laid out here may seem Malthusian in nature, there are important differences. A critical piece of the Malthusian view is unbridled population growth, which Malthus argued would consume all wealth accrued via technological change. However, population is not as critical to commodity demand growth as individual demand for meat and how it is linked to income growth in rapidly ascending countries like China. The Malthusian view also predicts subsistence living on brink of survival for the great majority of the population. But here the most dismal possible outcomes are unlikely to spread beyond the least developed countries with per-capita incomes well beyond subsistence. One possible scenario is that the today's poorest countries catch up with developed nations before any negative climate change impacts might begin to set in. This seems only likely if these countries switch from being food importers to food exporters. Rising commodity prices would then make them richer. We might expect even higher commodity price increases, high enough to induce significant substitution away from meat and toward less resource consumptive grains and vegetables. Severe climate-change impacts could still have significant welfare implications, because high prices would make agriculture a larger share of output. But widespread famine would seem unlikely given enough substitution away from meat toward grains and vegetables.

There are, of course, also good reasons to be optimistic. Emerging biotechnologies may bring about a second green revolution that facilitates even greater crop yield growth. Monsanto, the leading agricultural biotech company, has recently begun marketing "drought tolerant" corn varieties that may prove to be more heat tolerance than corn grown since 1950. New genetically engineered strains of wheat, rice, and soybeans might facilitate more yield growth for these crops too. Moreover, the prospect of adverse consequences could provide increasingly powerful incentives (i.e., through higher prices) for innovations that may aid or facilitate adaptation. There should be time for such adaptations, because climate impacts are likely to be realized gradually over many decades. Compelling research showing the potentially devastating impacts of climate change serves to further enhance the incentive innovate before any devastating impacts might be realized.

References

Arrow, Kenneth J., 'Global Climate Change: A Challenge to Policy,' *The Economists' Voice*, 2007, 4(3), 1–5.

Ashenfelter, Orley and Karl Storchmann, 'Using a Hedonic Model of Solar Radiation to Assess the Economic Effect of Climate Change: The Case of Mosel Valley Vineyards,' *NBER Working Paper 12380*, July 2006.

Cassman, Kenneth G., 'Ecological Intensification of Cereal Production Systems: Yield Potential, Soil Quality, and Precision Agriculture,' *Proceedings of the National Academy of Sciences*, May 1999, 96(11), 5952–5959.

Darwin, Roy, 'A Farmer's View of the Ricardian Approach to Measuring Agricultural Effects of Climate Change,' *Climatic Change*, March–April 1999, 41, 371–411.

Deschênes, Olivier and Michael Greenstone, 'The Economic Impacts of Climate Change: Evidence from Agricultural Output and Random Fluctuations in Weather,' *American Economic Review*, March 2007, 97(1), 354–385.

Fisher, Anthony C., W. Michael Hanemann, Michael J. Roberts, and Wolfram Schlenker, 'Potential Impacts of Climate Change on Crop Yields and Land Values in U.S. Agriculture: Negative, Significant, and Robust,' *Working Paper*, 2007.

Food and Agriculture Organization, 'FAO Statistical Year Book 2005–2006,' http://www.fao.org/statistics/yearbook/vol 1 1/index.asp, 2007.

Food and Agriculture Organization, http://faostat.fao.org/site/567/default.aspx, 2008.

Foreign Agricultural Service, 'Downloadable Data Sets,' http://www.fas.usda.gov/psdonline/psdDownload.aspx, 2008.

Gerbens-Leenes, P.W., S. Nonhebela, and W.P.M.F. Ivens, 'A Method to Determine Land Requirements Relating to Food Consumption Patterns,' *Agriculture, Ecosystems & Environment*, June 2002, 90(1), 47–58.

Johnson, D. Gale, 'Agriculture and the Wealth of Nations,' *American Economic Review*, May 1997, 87(2), 1–12.

Johnson, D. Gale, 'Population, Food, and Knowledge,' *American Economic Review*, March 2000, 90(1), 1–14.

Kelly, David L., Charles D. Kolstad, and Glenn T. Mitchell, 'Adjustment Costs from Environmental Change,' *Journal of Environmental Economics and Management*, November 2005, 50(3), 468–495.

Leibtag, Ephraim, 'Corn Prices Near Record High, But What About Food Costs?' *Amber Waves*, February 2008, 6(1), 10–15.

Lobell, David B., Marshall B. Burke, Claudia Tebaldi, Michael D. Mastrandrea, Walter P. Falcon, and Rosamond L. Naylor, 'Prioritizing Climate Change Adaptation Needs for Food Security in 2030,' *Science*, February 1 2008, 319, 607–610.

McCracken, Vicki A. and Jon A. Brandt, 'Household Consumption of Food-Away-from-Home: Total Expenditure and by Type of Food Facility,' *American Journal of Agricultural Economics*, May 1987, 69(2), 274–284.

Mendelsohn, Robert, William D. Nordhaus, and Daigee Shaw, 'The Impact of Global Warming on Agriculture: A Ricardian Analysis,' *American Economic Review*, September 1994, 84(4), 753–771.

National Agricultural Statistics Service, 'QuickStats—County Data—Crops,' http://www.nass.usda.gov/QuickStats/Create County All.jsp, 2008a.

National Agricultural Statistics Service, 'QuickStats—US and All States Data—Crops,' http://www.nass.usda.gov/QuickStats/Create Federal All.jsp, 2008b.

New York Times, 'A Drought in Australia, a Global Shortage of Rice,' April 17, 2008a.

New York Times, 'A Global Need for Grain That Farms Can't Fill,' March 9, 2008b.

New York Times, 'A New, Global Oil Quandary: Costly Fuel Means Costly Calories,' January 19, 2008c.

Roberts, Michael J. and Wolfram Schlenker, 'World Supply and Demand of Food Commodity Calories,' *American Journal of Agricultural Economics*, December 2009, 91(5).

Rosenzweig, Cynthia and Daniel Hillel, *Climate Change and the Global Harvest*, New York: Oxford University Press, 1998.

Schlenker, Wolfram and Michael J. Roberts, 'Non-linear Effects of Weather on Corn Yields,' *Review of Agricultural Economics*, 2006, 28(3), 391–398.

Schlenker, Wolfram and Michael J. Roberts, 'Nonlinear Temperature Effects indicate Severe Damages to U.S. Crop Yields under Climate Change,' *Proceedings of the National Academy of Sciences*, 2009, doi: 10.1073/pnas.0906865106.

Schlenker, Wolfram, W. Michael Hanemann, and Anthony C. Fisher, 'The Impact of Global Warming on U.S. Agriculture: An Econometric Analysis of Optimal Growing Conditions,' *Review of Economics and Statistics*, February 2006, 88(1), 113–125.

Searchinger, Timothy, Ralph Heimlich, R.A. Houghton, Fengxia Dong, Amani Elobeid, Jacinto Fabiosa, Simla Tokgoz, Dermot Hayes, and Tun-Hsiang Yu, 'Use of U.S. Croplands for Biofuels Increases Greenhouse Gases Through Emissions from Land-Use Change,' *Science*, 29 February 2008, 319(5867), 1238–1240.

Timmins, Christopher, 'Endogenous Land Use and the Ricardian Valuation of Climate Change,' *Environmental and Resource Economics*, January 2006, 33(1), 119–142.

Williamson, Lucille and Paul Williamson, 'What We Eat,' *Journal of Farm Economics*, August 1942, 24(3), 698–703.

3
Wealth, Saving and Sustainability

Kirk Hamilton

Deriving conceptually sound and useful indicators of sustainable development has been a challenge, not least because sustainability is inherently a concern about the future. Yet the question of indicators is key: without some means to quantify progress towards sustainability, all of the policy commitments by governments and institutions to achieving sustainable development risk becoming empty promises.

Pearce and Atkinson (1993) pointed the way on measurement. Their argument was largely intuitive – if gross saving in an economy is less than the combined value of depreciation of produced capital and depletion of natural resources, then future well-being must be at risk. They assembled empirical estimates of the value of depreciation, depletion and damage to the environment to show that many countries were apparently on an unsustainable path in the 1980s.

The literature on measuring sustainable development overlaps with the literature on the treatment of the environment and natural resources within the System of National Accounts (SNA). This is important because the SNA has a substantially incomplete treatment of resource and environmental issues. Because depletion and damage to the environment is ignored in the SNA measure of income, decisions to exploit natural assets are captured only partially – in the most extreme cases, asset liquidation gets accounted as income generation within the SNA.

An important strand of the research on measuring sustainable development since Pearce and Atkinson (1993) has treated the question more formally, working within the framework of growth theory. But the linkage between growth theory and national accounting is an intimate one, as the seminal paper by Weitzman (1976) has established. What follows, therefore, works within the growth-theoretic foundations of national accounting to highlight advances in the measurement of sustainable development.

We first present some general ideas regarding wealth and social welfare. Then the question of saving, the change in real wealth, is taken up in two sections dealing with optimal and non-optimal economies. Empirical estimates of saving and wealth from the World Bank are highlighted. And finally some issues and challenges are considered in the concluding section.

Wealth and social welfare

When economists speak of 'social welfare' they are explicitly including an inter-temporal dimension. The issue, of course, is that measuring current well-being does not tell you whether this well-being can be sustained in the future. To take a concrete example, during the first half of the 1980s fish catch in Mauritania grew strongly from around 20,000 tons in 1980 to nearly 90,000 tons in 1987. Fisheries were expected to provide a key source of growth, generating jobs, foreign exchange earnings and budget revenues. But the fishery collapsed from over-exploitation, with long-term consequences for growth – exports of goods and services grew at a real 7.5 per cent per year over 1980–87, but shrank by −2.3 per cent per year from 1987 to 2000. The well-being of Mauritanians benefiting from the export fishery could not be sustained. If economists had been measuring the total wealth of Mauritania, then the impending collapse could have been foreseen in the asset account for fish stocks.

The fact that income or consumption does not have a direct welfare interpretation was highlighted in a seminal paper by Samuelson (1961), who argued that the choice of a welfare measure has to be made 'in the space of all present and future consumption...the only valid approximation to a measure of welfare comes from computing *wealth-like* magnitudes not income magnitudes' (Samuelson 1961, pp. 50–57). Irving Fisher (1906) provided the original insight that the most complete measure of current wealth should be the present value of future consumption.[1] Fisher identified three types of assets: immovable wealth, comprising land and the fixed structures upon it; movable assets or commodities; and human beings.

In a recent paper Hamilton and Hartwick (2005) make these notions explicit in a competitive economy with a constant returns to scale production technology.[2] Total wealth W is defined as the sum of asset values,

$$W = \sum_{i=1}^{N} p_i K_i \tag{1}$$

Here the K_i are the stocks of assets in the economy, and the p_i are their shadow prices. To measure sustainability it is important that the wealth measure span as wide a range of assets as possible, including assets with negative shadow prices such as pollution stocks. Hamilton and Hartwick show that for interest rate r and consumption C,

$$W = \sum_{i=1}^{N} p_i K_i = \int_{t}^{\infty} C(s) \cdot \exp\left(-\int_{t}^{s} r(\tau)\, d\tau\right) ds \tag{2}$$

This is just what Fisher made explicit: total wealth is equal to the present value of future consumption, which in turn corresponds to Samuelson's notion of total wealth as a measure of social welfare.

Saving in optimal economies

If wealth measures social welfare, then changes in wealth should tell us about changes in social welfare and – as will be made explicit below – sustainability. Hamilton and Clemens (1999) grounded the insights of Pearce and Atkinson (1993) in the theory of optimal growth. They show that genuine saving G, defined as the change in real asset values,

$$G = \sum_{i=1}^{N} p_i \dot{K}_i \tag{3}$$

is equal to the change in social welfare in an optimal economy. That is, for utility U, social welfare V (here measured in utility units rather than consumption units), marginal utility of consumption λ and constant pure rate of time preference ρ:

$$G = \lambda^{-1} \dot{V} \quad \text{for} \tag{4}$$

$$V = \int_{t}^{\infty} U(C, \ldots) \cdot e^{-\rho(s-t)}\, ds. \tag{5}$$

This says that social welfare is equal to the present value of utility, and that genuine saving is equal to the instantaneous change in social welfare measured in dollars.[3] The utility function can include consumption C and any other set of goods and bads to which people attribute value.

Hamilton and Clemens (1999) go on to show that negative levels of genuine saving must imply that future levels of utility over some period of time are lower than current levels – that is, negative genuine saving implies unsustainability (the converse does not hold – Pezzey 2004 derives this result in a more general framework). This result links general notions of social welfare to the question of sustainable development.

Because this result holds in an optimal economy, its usefulness can be questioned. Hamilton and Hartwick (2005) provide a partial response in the context of the Dasgupta-Heal economy characterized by fixed technology, constant returns to scale and an exhaustible resource that is essential for production.

A key result in Dasgupta and Heal (1979) is that the optimal path for the simple exhaustible resource economy is not a sustainable path – Figure 3.1 shows a typical optimal path, where consumption rises, reaches a peak and then falls towards 0 asymptotically. This is driven by the fact that the marginal product of capital falls along the optimal path, until it is actually lower than the pure rate of time preference.

Hamilton and Hartwick (2005) establish that consumption and genuine saving are related in the following way along the optimal path:

$$\dot{C} = rG - \dot{G} \tag{6}$$

which can be solved to yield,

$$G = \int_{t}^{\infty} \dot{C}(s) \cdot \exp\left(-\int_{t}^{s} r(\tau)\, d\tau\right) ds \tag{7}$$

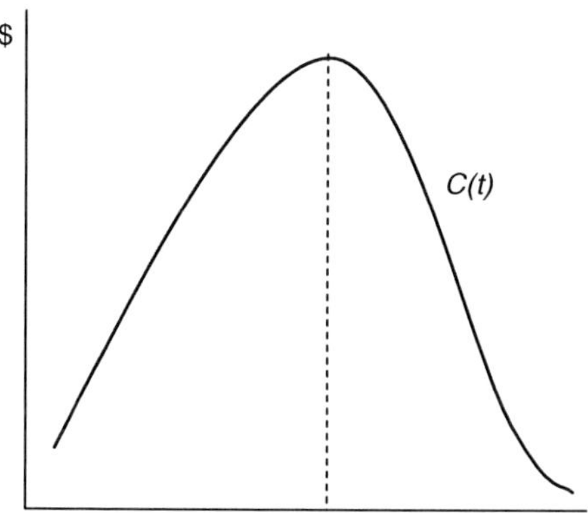

Figure 3.1   Consumption path in the optimal Dasgupta-Heal economy

This says that genuine saving is equal to the present value of future changes in consumption, a result derived in a more general setting in Dasgupta (2001, Chapter 9 Appendix A.7).

From expression (7) it is straightforward to see that if consumption is initially rising along the development path, as shown in Figure 3.1, then genuine saving will turn negative before the peak and subsequent decline in consumption is reached. Genuine saving, therefore, gives 'early warning' that a decline into unsustainability is impending. A hypothetical myopic social planner who is ensuring that the path for the economy is optimal would see that a decline is imminent and could opt for another policy to guide the economy.

Rules for sustainability in competitive economies

From a policy perspective the interesting question with regard to sustainable development is whether there are any general policy prescriptions which will ensure sustainability. For this it is necessary to look at non-optimal economies, because rules for sustainability will in most instances conflict with the necessary conditions for optimal growth – this simply reflects the divergence between sustainability and optimal growth as social objectives.

Dasgupta and Mäler (2000) offer one solution to this problem by looking at non-optimal economies which are driven by an allocation mechanism which determines the path of all future stocks and flows in the economy. For a suitable definition of the shadow prices in the economy, Dasgupta and Mäler show that expression (4) will hold – genuine saving in the non-optimal economy is proportional to the change in social welfare.

One alternative to presuming either optimality or a full allocation mechanism is to assume that the economy is competitive – roughly speaking, this implies that producers maximize profits over time, while households maximize their utility. This is the key assumption in Hamilton and Withagen (2007), who derive a generalization of expression (6) in competitive economies:

$$\dot{U} = \lambda G \left(r - \dot{G}/G \right) \tag{8}$$

This relates the change in utility to the difference between the interest rate and the growth rate of genuine saving and is the basis for the following general rule.

Rule for sustainability: In a competitive economy, a policy that ensures that $G > 0$ and $\dot{G}/G < r$ at each point in time in the future will ensure that the economy is sustainable.

Hamilton and Withagen (2007) go on to show that if the pure rate of time preference ρ is constant, then following this rule for sustainability into the indefinite future will also ensure that social welfare V is also increasing at each point in time.

Special cases of the rule for sustainability

The Hartwick rule. Perhaps the most famous rule for sustainability is that of Hartwick (1977), who shows that if genuine saving is equal to 0 at each point in time into the indefinite future, then utility will be constant. This, it can be seen, is a special case of the general rule for sustainability specified by Hamilton and Withagen (and foreshadowed in Hamilton and Hartwick 2005). Hartwick (1977) shows that the zero genuine saving rule is feasible in the Dasgupta-Heal economy if (i) the Hotelling rule holds, and (ii) the production technology is Cobb-Douglas with $\beta < \alpha$, where α and β are the elasticities of output with respect to produced capital and natural resources respectively.

Constant genuine saving rate. Hamilton and Withagen (2007) show that if F is production, R is resource extraction and p its shadow price, then

$$G = \dot{K} - pR = \gamma F \quad \text{for constant } \gamma \text{ satisfying} \quad 0 < \gamma < \alpha - \beta \quad (9)$$

is a feasible policy rule for sustainability, yielding unbounded consumption in the competitive Dasgupta-Heal economy.[4] They go on to show that the growth rate of net income $(C + G)$ in this economy is given by

$$\frac{\frac{d}{dt}(C+G)}{(C+G)} = \gamma \cdot \frac{r}{\alpha}. \quad (10)$$

For a 'typical' developing country where resources are perhaps 30 per cent of production (so $\alpha = 0.7$) and the rate of return on capital r is 7 per cent, this expression implies that each increase in genuine saving by 10 per cent of gross production will yield a 1 per cent increase in the growth rate of net income.

Constant level of genuine saving. Hamilton et al. (2006) show that if genuine saving is held fixed at some constant level $\overline{G}$ satisfying $0 < \overline{G} < \alpha F(K_0, R(0))$, then consumption is again unbounded in the competitive Dasgupta-Heal economy.

Testing the link between genuine saving and social welfare

Economic theory suggests that saving today and changes in future consumption should be linked. Ferreira and Vincent (2005) use World Bank historical data on consumption and genuine saving to test a basic proposition linking current saving to future welfare. They start with a result from Weitzman (1976): if the economy is optimal and the interest rate is constant then,

$$G = r \int_t^\infty C(s)e^{-r(s-t)}ds - C$$

Genuine saving is equal to the difference between a particular weighted average of future consumption and current consumption. This relationship is tested econometrically using per capita data from 1970 to 2000. Ferreira and Vincent find that the relationship holds best for non-OECD countries, and that there is a better fit as more stringent measures of saving are tested, that is, when going from gross saving to net saving to genuine saving (but excluding the adjustment for investment in human capital, which performs very badly).

This empirical test of genuine saving has been deepened by Ferreira et al. (2008), who analyze the wealth-diluting impact of population growth on measures of net saving per capita. Using a more robust model than Ferreira and Vincent (2005), Ferreira et al. show that only when net saving per capita is adjusted to reflect depletion of natural resources and wealth dilution is saving per capita correlated with changes in future well-being.

Empirical estimates of wealth and saving

The foregoing theory suggests that total wealth is useful for measuring social welfare, while changes in real wealth (genuine saving) can measure growth in social welfare and provide the basis for rules for sustainable development. This section of the chapter looks at the empirical evidence on wealth and saving as presented in *Where Is the Wealth of Nations?* (World Bank 2006).

This publication presents estimates of total wealth for over 100 countries by estimating the present value of future consumption over one generation (25 years), as derived from expression (2). Independent estimates of tangible assets (produced capital and stocks of natural resources including agricultural land, forests, minerals and energy) are then derived from national accounts data and data on physical stocks

Table 3.1 Estimates of total wealth by region and income group in 2000 (US$ per capita and %)

Group	Dollars per capita				Share of total wealth (%)		
	Total Wealth	Natural Capital	Prod. Capital	Intang. Capital	Natural Capital	Prod. Capital	Intang. Capital
Lat. Am. and Carib.	69,145	7,018	10,677	51,451	10	15	74
Sub-Saharan Africa	13,631	1,816	1,628	10,187	13	12	75
South Asia	6,906	1,749	1,115	4,043	25	16	59
East Asia and Pacif.	11,958	2,511	3,189	6,258	21	27	52
Mid. East and N. Africa	23,920	2,764	4,075	17,080	12	17	71
Eur. and Central Asia	41,964	3,795	8,446	29,722	9	20	71
Low Income	7,532	1,925	1,174	4,434	26	16	59
Lower Middle Income	22,674	2,970	4,187	15,517	13	18	68
Upper Middle Income	76,538	8,706	16,831	51,001	11	22	67
High Income OECD	439,063	9,531	76,193	353,339	2	17	80
World (excl. oil)	*95,860*	*4,011*	*16,850*	*74,998*	*1*	*17*	*82*
Oil exporters	22,952	12,656	7,937	2,359	55	35	10
World	90,210	4,681	16,160	69,369	5	18	77

Source: World Bank 2006.

and flows, prices and extraction/harvest costs for natural resources. The difference between total wealth and tangible wealth is derived residually and termed 'intangible capital' – Chapter 7 in *Where Is the Wealth of Nations?* shows that 90 per cent of the variation in intangible capital across countries can be explained by human capital and institutional quality as measured by an index of rule of law.

Table 3.1 summarizes the wealth estimation results by region, income group and for the world as a whole. High energy and mineral exporters are treated as a separate group (denoted 'oil exporters') because of their unique characteristics.

As Table 3.1 and Figure 3.2 below demonstrate, intangible capital is the preponderant form of wealth, an insight that goes back to the very origins of economic thinking.[5] The share of intangible capital rises across income classes, as expected.

The world's poorest countries – particularly in South and East Asia – depend heavily on natural resources (column 6 in Table 3.1). For low-income countries overall, natural resources constitute 26 per cent of total wealth, a share that is larger than produced capital. The natural resource share falls to 2 per cent of total wealth in high-income countries, but this is a fall in relative terms – figures in *Where Is the*

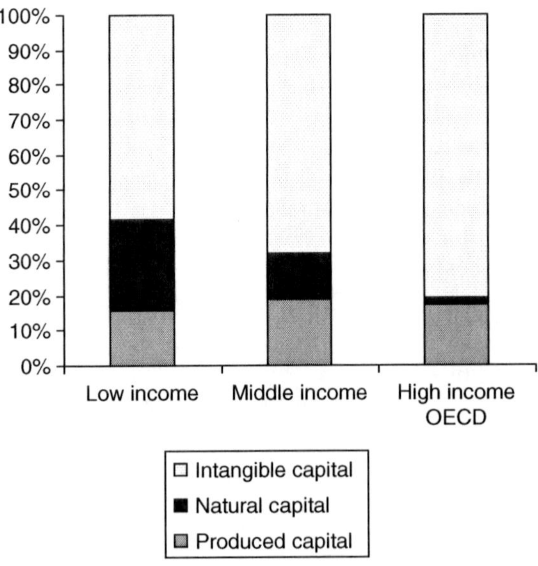

Figure 3.2   Composition of wealth by income
Source: World Bank 2006.

Wealth of Nations? show that the total value of natural capital per person actually rises with income.

The oil exporters stand out as a special case in Table 3.1, with only 10 per cent of total wealth composed of intangible capital. As argued in *Where Is the Wealth of Nations?* this almost certainly reflects the low returns on all assets that characterize these economies – resource rents of more than 20 per cent of gross national income (GNI) (in some cases much more) are highly distortionary. The 'resource curse' literature explores these issues more fully.

Saving estimates

Figure 3.3 shows the steps in calculating genuine saving for Bolivia, one of the poorest countries in Latin America, with GDP per capita below $1,000. Bolivia is endowed with a wealth of natural resources, including minerals, oil and huge deposits of natural gas discovered at the end of the 1990s.

The first column in Figure 3.3 shows the traditional measure of gross national saving in Bolivia, 12 per cent of GNI in 2003. Deducting the depreciation of produced capital reveals a much lower net saving rate, less than 3 per cent. Investments in education are estimated to be around 5 per cent of GNI, bringing the saving rate up to nearly 8 per cent as shown by the third column in Figure 3.3. Following this,

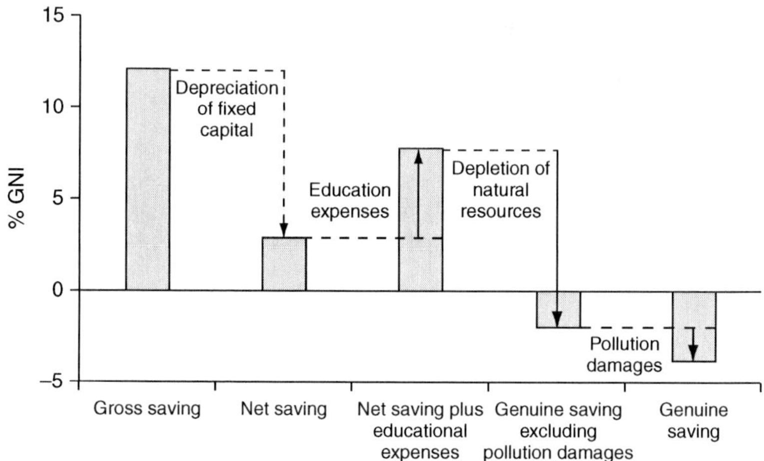

Figure 3.3 Genuine saving in Bolivia (2003)
Source: World Bank 2006.

adjustments are made for depletion of natural resources. Resource rents from Bolivia's extraction of oil and gas are deducted, as well as the rents from gold, silver, lead, zinc and tin. Depletion of energy, metals and minerals amounts to over 9 per cent of GNI. As a result of these deductions for resource depletion, Bolivia's genuine saving rate is negative. Finally, the deduction for pollution damages leads to a bottom-line estimate of Bolivia's genuine saving rate of −3.8 per cent of GNI. Bolivia is currently on an unsustainable development path.

As Figure 3.4 shows, aggregate savings for the developing regions of the world display distinctive levels and trends.

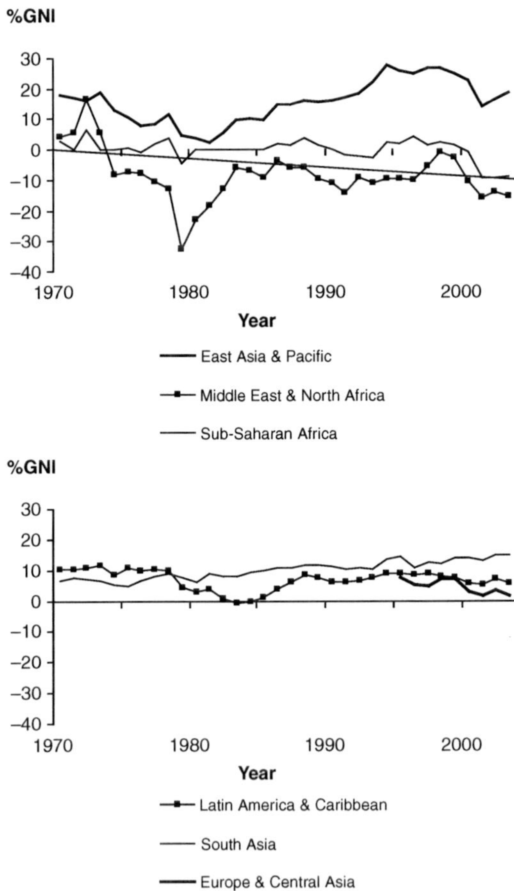

Figure 3.4   Trends in genuine saving by region
Source: World Bank 2006.

The Middle East and North Africa stands out for its consistently negative saving rate, reflecting high dependence on petroleum extraction. Regional genuine saving rates are highly sensitive to changes in world oil prices. This is clearly shown in Figure 3.3 – genuine saving rates dropped in 1979, largely owing to the consumption of sharply increased oil rents following the Iranian revolution.

East Asia and Pacific, and, to some extent, South Asia, stand in stark contrast, with recent aggregate genuine saving figures nearing 30 per cent, driven largely by China. The boom in economic performance from the second half of the 1980s until the Asian financial crisis in 1997 is reflected in the genuine saving numbers, largely driven by increases in gross national saving.

Genuine saving rates have been hovering around zero in Sub-Saharan Africa. Positive saving in countries such as Kenya, Tanzania and South Africa is offset by strongly negative genuine saving rates in resource-dependent countries such as Nigeria and Angola, which have genuine saving rates of −30 per cent in 2003.

Latin American genuine savings rate have remained fairly constant throughout the 1990s. The large economies in the region, Mexico and Brazil, have positive genuine saving rates in excess of 5 per cent. However, like many oil producers, Venezuela's genuine saving rate has been persistently negative since the late 1970s.

This is the broad picture that emerges from the cross-country analysis of wealth and savings: Intangible wealth, including human and institutional capital, is the most important share of wealth, and this share increases with income; in low-income countries natural resources are the next most important share of wealth; genuine saving rates are negative in many of the most resource-dependent economies, and have been effectively zero in Sub-Saharan Africa over the last three decades.

Challenges, future directions and conclusions

These approaches to extending the national accounts, and the models that underpin them, are agnostic on the question of the degree of substitutability between different assets, in particular between produced and natural assets. An important strand of the sustainability literature, dating back to Pearce et al. (1989), looks at the question of *strong* versus *weak* sustainability. Weak sustainability assumes that there are no fundamental constraints on substitutability. If, however, some amount of nature must be conserved in order to sustain utility – the strong sustainability assumption – then these saving models need to be modified to incorporate the shadow price of the sustainability constraint.

A formal approach to the strong versus weak sustainability problem has been explored in the Hartwick rule literature. Dasgupta and Heal (1979) and Hamilton (1995) show that if the elasticity of substitution between produced capital and natural resources is less than 1, then the Hartwick rule is not feasible – eventually production and consumption must fall, implying that the economy is not sustainable under the rule.

The question of environmental thresholds is potentially important in measuring sustainable development. Crossing certain physical boundaries may produce catastrophic results, such as the loss of the Greenland and Antarctic ice sheets as a result of global warming, or the death of plankton in the ocean as a result of ozone layer destruction. In environmental economic terms we may think of a threshold as a point where the marginal damage curve is unbounded. As long as marginal damages are smooth as a threshold is approached, the saving indicator will give correct signals concerning sustainability, since approaching the threshold will eventually result in negative savings. If the marginal damage curve is not smooth and becomes vertical at the threshold, then the saving rule may not indicate unsustainability as the threshold is approached. There is clearly an important question of the science of threshold problems, since we do not know *a priori* what the shape of the marginal damage curve is for many important problems.[6]

There are important questions about the measurement of shadow prices in this accounting work, particularly for natural resources. If we assume that world prices for resources do reflect scarcities and are therefore relatively undistorted, then the derived shadow prices should be a reasonable reflection of the social costs associated with resource extraction. Dasgupta and Mäler (2000) perhaps point a way forward with their suggestion that 'accounting' prices should measure the marginal contribution of individual assets to social welfare.

A related valuation problem concerns the treatment of capital gains in valuing the depletion of exhaustible resources. Standard approaches such as the 'El Serafy' method (El Serafy 1989) implicitly include holding gains in the derivation of values of depletion, while the theory embodied in expressions (3) and (4) above suggests that only changes in real wealth can indicate changes in social welfare.

These questions suggest a further range of issues where future research will be required, including:

- Identifying thresholds and non-linearities in the natural world that may not be captured in any simple way in measures of genuine saving.
- Inventorying and valuing the environmental services that underpin so much economic activity. While many of these values are captured indirectly in other asset values – the value of farmland includes the value of pollination services, for example – the fact that there is no explicit valuation means that there are opportunities for unpleasant policy surprises. Robust valuation of truly difficult-to-value assets such as biodiversity is also a priority.
- Estimating elasticities of substitution for resources. The availability of databases of natural resource stocks and flows, in quantity and value terms, means that there should be more scope for exploring this important question – World Bank (2006, Chapter 8) estimates the elasticity of substitution between land and fixed capital to be close to 1, an important result.

The foregoing theory tells us that genuine saving is the correct measure of the change in social welfare and that negative genuine saving indicates that an economy is on an unsustainable path. This measure of genuine saving provides the basis for a general rule for sustainable development: ensure that saving is positive and growing at a rate less than the interest rate. The Hartwick rule is a special case of this more general rule for sustainability. The presentation of the empirical estimates reminds us that, for both conceptual and practical reasons, our measures of wealth and saving are incomplete, limiting the precision of our measures of sustainability. This fact gives rise to a rich research agenda going forward.

Notes

1. Fisher's argument was motivated by the need to find a measure of comprehensive wealth. This led to the intuition that the value of an asset is the capitalization of the stream of future services expected to be produced by the asset.
2. See Dixit et al. (1980) for details on a competitive economy. A key consequence of assuming a competitive economy is that shadow prices should be dynamically efficient, for example, the Hotelling rule for the scarcity rents on exhaustible resources. While Hamilton and Hartwick (2005) actually establish

their result in an optimal economy, it holds in a competitive economy as well.

3. This result is foreshadowed in Aronsson et al. (1997, expression 6.18) who show that net saving measured in utils is equal to the present value of changes in utility for a general (possibly time-varying) pure rate of time preference. Similarly, Asheim and Weitzman (2001) show that growth in real Net National Product (where prices are deflated by a Divisia index of consumption prices) indicates the change in social welfare in the economy.

4. Asheim and Buchholz (2004) derive a similar result for a constant gross saving rate rule.

5. In *An Inquiry into the Nature and Causes of the Wealth of Nations*, Adam Smith wrote: 'The annual labour of every nation is the fund which originally supplies it with all the necessaries and conveniences of life which it annually consumes.' Smith recognized 'the skill, dexterity, and judgment with which [...] labour is generally applied' as a precondition for generating supply 'whatever be the soil, climate, or extent of territory of any particular nation'.

6. See also Pearce et al. (1996).

References

Aronsson, T., P.-O. Johansson and K.-G. Löfgren, 1997. *Welfare Measurement, Sustainability and Green National Accounting: A Growth Theoretical Approach*, Cheltenham: Edward Elgar.

Asheim, G. and W. Buchholz, 2004. A General Approach to Welfare Measurement Through National Income Accounting. *Scandinavian Journal of Economics* 106, 361–384.

Asheim, G.B. and M.L. Weitzman, 2001. Does NNP Growth Indicate Welfare Improvement? *Economics Letters* 73:2, 233–239.

Dasgupta, P., 2001. *Human Well-Being and the Natural Environment*, Oxford: Oxford University Press.

Dasgupta, P. and G. Heal, 1979. *Economic Theory and Exhaustible Resources*, Cambridge: Cambridge University Press.

Dasgupta, P. and K.-G. Mäler, 2000. Net National Product, Wealth, and Social Well-Being. *Environment and Development Economics* 5, Parts 1&2, 69–93, February and May 2000.

Dixit, A., P. Hammond and M. Höel, 1980. On Hartwick's Rule for Regular Maximin Paths of Capital Accumulation and Resource Depletion. *Review of Economic Studies* XLVII, 551–556.

El Serafy, S., 1989. The Proper Calculation of Income from Depletable Natural Resources, in Y. Ahmad, S. El Serafy and E. Lutz (eds), *Environmental Accounting for Sustainable Development*, Washington DC: The World Bank.

Ferreira, S. and J. Vincent, 2005. Genuine Savings: Leading Indicator of Sustainable Development? *Economic Development and Cultural Change* 53:3, 737–754.

Ferreira, S., K. Hamilton and J. Vincent, 2008. Comprehensive Wealth and Future Consumption: Accounting for Population Growth. *World Bank Economic Review* 22:2, 233–248.

Fisher, I., 1906. *The Nature of Capital and Income*, New York: Macmillan.

Hamilton, K., 1995. Sustainable Development, the Hartwick Rule and Optimal Growth. *Environmental and Resource Economics* 5:4, 393–411.

Hamilton, K. and M. Clemens, 1999. Genuine Savings Rates in Developing Countries. *World Bank Economic Review* 13:2, 333–356.

Hamilton, K. and J.M. Hartwick, 2005. Investing Exhaustible Resource Rents and the Path of Consumption. *Canadian Journal of Economics* 38:2, 615–621.

Hamilton, K. and C. Withagen, 2007. Savings Growth and the Path of Utility. *Canadian Journal of Economics* 40:2, 703–713.

Hamilton, K., G. Ruta and L. Tajibaeva, 2006. Capital Accumulation and Resource Depletion: A Hartwick Rule Counterfactual. *Environmental and Resource Economics* 34, 517–533.

Hartwick, J.M., 1977. Intergenerational Equity and the Investing of Rents from Exhaustible Resources. *American Economic Review* 67:5, 972–974.

Pearce, D.W. and G. Atkinson, 1993. Capital Theory and the Measurement of Sustainable Development: An Indicator of Weak Sustainability. *Ecological Economics* 8, 103–108.

Pearce, D.W., A. Markandya and E.B. Barbier, 1989. *Blueprint For a Green Economy*, London: Earthscan.

Pearce, D.W., K. Hamilton and G. Atkinson, 1996. Measuring Sustainable Development: Progress on Indicators. *Environment and Development Economics* 1, 85–101.

Pezzey, J., 2004. One-sided Sustainability Tests with Amenities, and Changes in Technology, Trade and Population. *Journal of Environmental Economics and Management* 48:1, 613–631.

Samuelson, P., 1961. The Evaluation of 'Social Income': Capital Formation and Wealth, in F.A. Lutz and D.C. Hague (eds), *The Theory of Capital*, New York: St. Martin's Press.

Smith, Adam, 1776. *An Inquiry into the Nature and Causes of the Wealth of Nations*, Chicago: University of Chicago Press, 1977.

Weitzman, M., 1976. On the Welfare Significance of National Product in a Dynamic Economy. *Quarterly Journal of Economics* 90:1, 156–162.

World Bank, 2005. *World Development Indicators*, Washington: The World Bank.

World Bank, 2006. *Where Is the Wealth of Nations? Measuring Capital for the 21st Century*, Washington: The World Bank.

4

China, the US, and Sustainability: Perspectives Based on Comprehensive Wealth

Kenneth J. Arrow, Partha Dasgupta, Lawrence H. Goulder, Kevin Mumford, and Kirsten Oleson

I Introduction

Policy analysts and policy makers are keenly interested in whether the performance of national economies is consistent with some notion of "sustainability." This reflects growing concerns about environmental quality and about the depletion of oil reserves and other natural resource stocks. Economists and natural scientists have offered several notions of sustainability. An especially important notion—and the one on which this chapter focuses—is defined with reference to human well-being. This notion of sustainability is achieved if the current generation leaves the next one with the capacity to enjoy the same or higher quality of life. Standard measures in the national income accounts—such as changes in per-capita GDP—may offer hints of whether a nation meets this sustainability criterion, but as is well known these measures do not fully capture many important contributors to well-being, such as the changes in the stocks of natural capital or in environmental quality.

The issue of sustainability seems especially relevant to China today. Although estimates vary, per-capita GDP in China appears to have grown at an annual rate of over 8 percent over the past 15 years.[1] In terms of marketed goods and services, the nation appears to be making extremely good progress. At the same time, China has accomplished this GDP growth through significant reductions in its natural resource base. According to China's State Forestry Administration, itinerant farming has contributed to soil erosion on a large scale, with desert expanding at a rate of 10,400 square kilometers per year. China's cities rank among the world's worst for air pollution, and all of China's major waterways are classified as "severely polluted" by the World Resources Institute (1998). This loss of natural capital offsets the positive contribution to the productive base from investments in reproducible capital. As a result, it is not immediately clear whether the China's *overall* productive base is rising or even being maintained. As discussed below, the overall productive base is intimately connected to the ability of the nation to generate goods and services and thus maintain living standards—which is at the heart of our notion of sustainability. Furthermore, China's rapid GDP growth has come at considerable cost in terms of environmental quality. Is per-capita well-being sustainable, given the losses of natural capital and environmental quality?

The sustainability issue also applies to the US, but perhaps in a different way. A growing share of the US capital base is owned by foreigners. The sustainability of well-being to US residents is closely connected to the changes in per-capita wealth owned by these residents. Is per-capita wealth of US residents rising and, if so, at what rate?

This chapter addresses these and other questions. Our overall objective is to shed light on whether China and the US are meeting a sustainability criterion. It can be shown (e.g., Arrow et al. 2004; see below) that, under a wide set of circumstances, intergenerational well-being is sustainable during a period of time if and only if a comprehensive measure of wealth per capita is nondeclining during that same period. This comprehensive wealth measure encompasses a wider range of productive assets than those in traditional national accounts. It embraces not only reproducible capital but also human capital and many commercial forms of natural capital. In addition, the focus on wealth directs attention to the entire intertemporal stream of goods and services implied by today's assets, rather than the current flow of income.

This effort is in the general category of comprehensive wealth accounting. Some of the most important advances in such accounting have been made in recent years by Kirk Hamilton and his collaborators at the World Bank. Hamilton and Clemens (1999) explored whether comprehensive wealth is rising or falling in various developing countries. Arrow et al. (2004) built on the World Bank's framework by incorporating technological change and considering population growth. In *Where Is the Wealth of Nations?* (World Bank 2006), a World Bank team headed by Hamilton provides assessments of changes in comprehensive wealth for nearly every nation of the world.

The present chapter aims to advance comprehensive wealth accounting in several ways. First, we offer a more theoretically consistent approach to valuing natural resources. This includes attention to how future changes in natural resource prices can influence comprehensive wealth measured today. This is especially important in regard to reserves of crude petroleum. Second, we offer an improved approach to measuring changes in human capital. While prior work used education expenditure as a proxy for the change in human capital, we employ a measure based on estimates of changes in educational attainment. Third, we explicitly distinguish between domestic and foreign holdings of a nation's capital. Fourth, we introduce an improved treatment of changes in wealth connected with environmental damages associated with climate change.[2]

The chapter is organized as follows. The next section lays out the main elements of our analytical framework. Section III then applies the framework to examine the changes in per-capita comprehensive wealth in China and the US over the period 1995–2000. Section IV offers conclusions and suggests directions for future work.

II   Methodology

A   A sustainability criterion

Researchers have offered a great many definitions of sustainability, as evidenced by Pezzey's (1992) survey of the various notions. Our sustainability requirement focuses on intertemporal welfare. (See Arrow et al. 2004, pp. 150–154 for discussion and references.) According to this approach, the (intertemporal) welfare of any one generation is determined not merely by its utility for current consumption but also for the care it has for future generations. We let V denote intertemporal welfare. One possible expression for V is:

$$V(t) = \int_t^\infty e^{-\delta t} U[c(u)]du \qquad (1)$$

where t is time, δ is the subjective rate of discount of utility (time preference), U is satisfaction or *felicity* at any moment of time, and c is an aggregate vector of different kinds of consumption. The c vector includes not only marketed goods but also amenity values of natural resources, and various dimensions of health. The criterion of sustainability is that V is nondecreasing:

$$dV/dt \geq 0. \qquad (2)$$

The possibilities for consumption are determined by an economy's *productive base*, an index of the quantities available of a number of types of capital. The capital assets include (1) manufactured capital goods, referred to as "reproducible capital," (2) human capital, the productive capacity inhering in human beings and acquired through education,[3] and (3) various kinds of natural capital. Natural capital includes land and various mineral resources.

Production of new goods takes place according to a technology which relates the use of various forms of capital to outputs. For simplicity, and again in accord with standard models of economic growth, this analytical framework assumes there is one output, which can either be consumed or added to reproducible capital. Natural resources may be nonrenewable, as with minerals, or renewable, as with forests. In the former case, the stock of the natural resource in any period is reduced by the quantity extracted (the flow) in that period. In the latter, the stock is increased by its natural rate of growth as well as being reduced by the flow used. The rate of change in the stock of a particular kind of capital is called the *investment* in that kind of capital. Investment in nonrenewable natural resources is necessarily negative.

The output generated by the productive base divides between consumption goods and services and investment in reproducible capital. We assume that allocation rules (which may include functions of market prices) determine the allocation of output between consumption and investment, and that the allocation system is *autonomous*, by which we mean that V is not an explicit function of time. Hence the stocks of the different kinds of capital in the next period are determined by the stocks in the present period and the (fixed) allocation rules.

By proceeding from period to period this way, the entire future course of capital stocks and therefore of flows of investment (by following the allocation rules) is determined.[4] Given the stocks of the different kinds of capital, $K_i(i=1,\ldots n)$ at some time t, the values of K_t and consumption c are determined at all future times $u \geq t$. Hence $U[c(u)]$ is determined for all $u \geq t$, and, from (1), $V(t)$ is determined as well. Hence we can write:

$$V(t) = V[K_1(t),\ K_2(t),\ldots,\ K_n(t)]. \tag{3}$$

Therefore, from (2), sustainability requires that

$$dV/dt = \sum_{i=1}^{n} (\partial V/\partial K_i)(dK_i/dt) > 0. \tag{4}$$

The theory we are invoking here does not require that intergenerational well-being, V, has the functional form given in equation (1). Let $q_i \equiv \partial V/\partial K_i$ and $I_i \equiv dK_i/dt$. The variable q_i is the marginal contribution of the ith type of capital to intertemporal welfare, and thus may be thought of as the *shadow price* of that kind of capital. I_i is the time derivative of the capital stock K_i or, in the usual terminology, the *investment* in that capital. It follows that

$$dV/dt = \sum_{i=1}^{n} q_i I_i. \tag{5}$$

Thus, dV/dt is the value of the new investments in different kinds of capital evaluated at the shadow prices. This suggests an interpretation of dV/dt as the change in wealth evaluated at constant prices, that is, the change in real wealth. Since we are including all forms of wealth, including natural resources, we refer to this as the change in *comprehensive* wealth.[5] Hence, from (4), the criterion for sustainability is precisely that real wealth is increasing.

The shadow prices are the prices that would prevail if all commodities were traded in competitive markets and if there were perfect foresight. Thus the shadow price for a nonrenewable resource such as oil is the discounted value of future use. It is therefore the price at which the owner of the well would be indifferent between selling the oil now and holding it for future sale. More precisely, the shadow price is the difference between the sales price of the oil and the cost of its extraction; it is the price paid for the *scarcity* of the resource.

The shadow prices are stated above in units of utility per unit capital. In view of the arbitrariness in the choice of units for utility, it is useful to employ a different *numéraire*; a natural choice is the aggregate commodity which can be used for either consumption or reproducible capital. This is the same technique as is used in ordinary price indices. Let reproducible capital be given the index 1 in the enumeration of types of capital. Then define the shadow prices of the different kinds of capital measured in terms of reproducible capital,

$$p_i = q_i/q_1, \quad (i = 1, \ldots n), \tag{6}$$

and the change in comprehensive wealth in the same terms:

$$dW/dt = \sum_{i=1}^{n} p_i I_i \tag{7}$$

Here $p_1 = 1$, from (6). Hence sustainability requires that

$$dW/dt > 0. \tag{8}$$

The formalism used here permits a measure of comprehensive wealth, as well as of the change in comprehensive wealth. In the notation already used,

$$W = \sum_{i=1}^{n} p_i K_i. \tag{9}$$

This explicit measure of comprehensive wealth is designed to replace the rough approximations used in Arrow et al. (2004, Table 1 and Note, p. 163).

B Measuring investments and determining shadow prices

1 Natural capital

To value the changes in natural capital, we need to consider both the net investment (DK) and the shadow price to apply to that investment. The net investment in a nonrenewable resource is simply the negative of the amount used up. The shadow price is related to the rental value of the resource. As is well known since the classical analysis of Hotelling (see, e.g., Dasgupta and Heal 1979), in a competitive setting the rental value of a nonrenewable resource should rise at the rate of interest (the marginal productivity of capital). If we abstract from externalities associated with use of the resource, then the rental value will correspond to the resource's shadow price.

For renewable resources, such as forests, the shadow price is again the rental value (price less cost of cutting), but the net investment equals the increase in the forests because of natural growth and planting less the amount used up.

2 Capital gains in nonrenewable resources

To the extent that the rental value of a nonrenewable resource rises through time, owners of the resource stock should expect to receive capital gains. Similarly, future consumers should expect to pay higher real prices. Other things equal, this implies a reduction in real wealth. Thus the impacts on real wealth of a given nation's residents will depend on the extent to which the residents own (and sell) or consume (purchase) the resource in question. In the empirical application below, we account for these wealth impacts. It appears that these impacts have not been addressed in any of the prior literature.[6] It may be noted that in a closed economy there is no need to adjust wealth for capital gains or losses, since the future gains to owners will be exactly offset by the losses to future consumers

For each country, the capital gain is equal to the stock of the resource times the rate of increase of the shadow price (i.e., the rate of interest). Summing over all countries gives the total capital gains to that resource. The corresponding capital losses by purchasers must be equal to this sum. In principle, it should be allocated among individual countries in accordance with their future purchases of oil. In the empirical application below we have approximated by giving each country a capital loss equal to total capital losses to consumer times that country's share of current consumption.

3   Human capital

We follow the methods introduced by Klenow and Rodríguez-Clare (1997), which builds on the earlier work of Mincer. That is, it is assumed that education is taken to earn a market rate of interest for the period of education. Assuming, as a first approximation, a steady state, the amount of human capital per worker is proportional to $e^{\rho A}$, where ρ is the appropriate rate of interest (taken to be 8.5 percent per annum) and A is the average number of years of educational attainment.

The stock of human capital, then, is the human capital per worker multiplied by the number of workers. This quantity is adjusted for mortality during the working life.

We assume that the labor market is sufficiently competitive, and that the marginal productivity of human capital is equal to its shadow price and also equal to the real wage. Hence the shadow price of human capital is equal to the total real wage bill divided by the stock of human capital.

4   Technological change

In the presence of technological change, the rate of growth of wealth is increased beyond that indicated by the growth in the stocks of individual kinds of capital, as displayed in (8).[7]

We follow the treatment of Arrow et al. (2004, footnote 7, pp. 153–154), adjusted to a different specification of the production function. Arrow et al. assumed that output is a function of reproducible capital and labor, so the elasticity of output with respect to capital was assumed to be a constant α less than 1. We now follow Klenow and Rodríguez-Clare (1997, 2005) in making output a function of two kinds of capital, reproducible and human. Thus the elasticity of output with respect to all forms of capital is now one. Hence, from Arrow et al. (2004, footnote 7), where α is set equal to 1, the adjustment to the rate of growth of real wealth is obtained by adding the Hicks-neutral rate of technological progress to the rate of growth of the aggregate of other forms of capital.

5 Population

Again we follow the usage of Arrow et al. (2004, pp. 152–153). If population is changing, then the appropriate measure of sustainability is that real wealth per capita should be growing. From (9) and (10), the sustainability criterion for a changing population is that

$$(dW/dt)/W - \pi > 0, \tag{10}$$

where π is the rate of growth of population (assumed to be exogenous).

6 Climate change and other environmental externalities

Our aim is to subtract from growth in comprehensive wealth the damages caused to a country by anthropogenic climate warming and other pollution externalities. Our approach differs from that in Hamilton and Clemens (1999) and used in Arrow et al. (2004, Table 1, p. 163), which assume that the climate change damages to a given country depend entirely on that country's CO_2 emissions. In contrast, our approach considers global emissions (rather than just those of the US and China) over the time-interval of interest, calculates the estimated damages from these emissions (now and in the future), and attributes a share of the global damages to the US and China. The estimated damages are then subtracted from other investments in the calculation of comprehensive investment.[8]

III Empirical application

In this section we estimate comprehensive wealth for China and the US in 1995 and 2000. As mentioned, comprehensive wealth accounts for the values of natural, human, and reproducible capital.

Our empirical application proceeds in two main steps. First, we evaluate the levels and changes in various stocks of capital over the 1995–2000 time-interval. We then consider the change in wealth on a per-capita basis, and make an adjustment for technological change.

A   Levels and changes in capital stocks

1   *Natural capital*

Natural capital includes exhaustible energy and mineral resources as well as renewable forest and land resources. We focus on the economically most important types of natural capital, to the extent that data are available.

a. Oil and natural gas.   As indicated in Section II.B.1, for nonrenewable resources the appropriate price to apply to the change in the capital stock is the scarcity rent on the resource. For several nonrenewable resources—particularly those with remaining reserves large enough to last more than 100 years at current rates of extraction—the estimated reserves are so large as to make the scarcity rents negligible. We therefore ignored nonrenewable resources whose remaining reserves could provide for over 100 years of use at current extraction rates. We ignored hard and soft coal, bauxite, and iron ore for this reason. We focus instead on oil, natural gas, other metal and mineral resources, forests, and land.

Not all of the stock of oil or natural gas is close enough to the surface or in a form that it is likely to be extracted given current technology and prices. To measure this, petroleum engineers use two categories: proved reserves and unproved reserves. Proved reserves are the stock of the resource that is estimated to be commercially recoverable under the current economic conditions, technology, and government regulation. Unproved reserves are reserves that are unlikely to be commercially recoverable under current conditions.

In recent years, changing prices and new operating methods have allowed petroleum engineers to increase the stock of these energy resources that they characterize as proved reserves. In fact, from 1995 to 2000, the proved reserves of oil in the world increased by 8.7 percent to 1115.8 billion barrels (even after more than 130 billion barrels were extracted). The proved reserves of natural gas increased even more rapidly at 12.1 percent over this same period. These increases are not only the result of changing economic conditions; they also reflect annual discoveries of oil and natural gas.

To calculate consistently the changes in the resource base, we start with a recent estimate of the proved reserves for the resource and then to back out, using production data, the stock of the resource in prior years. This irons out the impact of new discoveries and emphasizes the idea that, whatever the true global stock of reserves, this stock is diminished by the amount of extraction. Thus, given the estimated stock of a nonrenewable resource at the end of year t, the stock at the end of year $t-1$ is given by

$$K_{t-1} = K_t + X_t \tag{11}$$

where K_t and X_t represent proven reserves and extraction, respectively, in period t. The 2004 proved reserves and production data for oil and natural gas was obtained from the *BP Statistical Review of World Energy* (2005). We take the 2004 proved reserve and then add the quantity produced during the year to calculate the 2003 stock. We repeat this method to calculate the stock in 1995 and 2000.

To value the stock of a particular resource, we use the average unit rent in 1995 and 2000 for each country. This is the difference between the average real price and the average real extraction cost which, as an approximation, we assume reflects the shadow value of the decline in the resource stock. We assume a constant world price during the period, measured as the real average of spot prices over 1995 to 2000. For oil we average the price of four types of crude (Dubai, Brent, Nigerian Forcados, and West Texas Intermediate) and for natural gas we average the price from four sources (US, UK, Japan, and European Union). The extraction costs, obtained from the World Bank (2005), are based on several different studies. For both energy resources, China's extraction costs are not given, so we use 80 percent of the US estimate for oil and the world average for natural gas.

b. Metals and minerals. We follow the same approach for measuring the stock of metal and mineral resources in each country. The stock of each resource in 2000 and the annual production volumes were assembled from various sources, including the World Bank's *Where Is the Wealth of Nations*, United States Geological Survey, and other sources. As was explained for oil and natural gas, the stock of the resource in the previous year is obtained by deducting the quantity produced during the year.

Average world market price and extraction costs for each resource in each country were obtained from World Bank data (2005). As described above, we use the difference over the period between the average real price and the average real extraction cost to calculate the shadow value (scarcity rent) of one unit of the stock. If for any year between 1995 and 2000 the world market price was below the country's extraction cost, we eliminated the metal or mineral from the calculus. Our assumption in these cases was again that the scarcity rent was negligible since the world price was below extraction cost. As a result, we eliminated gold, nickel, tin, silver, and zinc from the analysis. If, for all years in 1995–2000, price was greater than cost, we averaged the difference between annual real price and annual real extraction cost, adjusted to year 2000 dollars, to obtain our measure of the average unit rent for the period 1995–2000.

c. Forests.   While globally the area of forest cover continues to decline (mainly because of conversion to agriculture—see, FAO 2005), forest area and forest stock increased between 1995 and 2000 in both the US and China. This increase is largely due to afforestation on productive plantations. China and the US account for 42 percent of the world's area of productive plantations. Over the interval 2000–2005, China's increase in forest area was the largest in the world, increasing by 4058 hectares per year and dwarfing the gains in the US (the 4th largest net gainer at 159 hectares per year).

We obtained total cubic meters of commercially available forests from the Food and Agriculture Organization (FAO) Forestry Resources Assessment (FRA) (FAO 2005). These data include volumes of growing stock of forests and other wooded resources in 1990, 2000, and 2005, and designate the amount of total stock that is commercially available. The difference in stock from one year to the next is assumed to be "produced" (if negative) or "afforested" (if positive). Subject to all the caveats that can be justifiably raised regarding the comparability of cross-country statistics, the stock would appear to have increased in the US and in China during 1995–2000.

For the accounting (shadow) price on forests, we used the rental value. The rental value was calculated as the weighted average market price of the types of wood minus the extraction costs. Extraction costs, specific to each country, were obtained from the World Bank's Adjusted Net Savings data. The resulting accounting price is $52 per cubic meter for the US, and $30 per cubic meter for China.

Note that this differs significantly from the previous World Bank method, where all estimates of commercially valuable area, stock per hectare, and net annual increase were independently estimated. Because the volume of commercial stock was included in FAO *Forest Resources Assessment*, we eliminated these "judgment calls" (but were therefore forced to use region-specific information in some cases). Because of the recent reversal in the historical trend of deforestation in these two nations, we credited them with the value of its afforested stock. In future calculations we would like to include afforestation costs, which can be significant expenses.

d. Land. Values for land include nontimber forest resources, protected areas, cropland, and pastureland. We obtained these values using information from the World Bank, as presented in *Where Is the Wealth of Nations?* Briefly, the World Bank uses two studies to estimate the value of nontimber forest resources. One-tenth of forested area is considered "accessible" to these kinds of nonextractive activities, and is assigned a value of $190 per hectare in developed countries and $145 per hectare in developing nations. The value of cropland is set equal to the present discounted value of land rents, which are based on a percentage of estimated production revenue for an array of crops sold at world market prices. The total land rent is the area-weighted average of rents from major crops. Pastureland is valued as the opportunity cost of preserving land for grazing. Returns are calculated assuming a fixed proportion of value to output (returns are estimated at 45 percent of output), where output is based on production of beef, lamb, milk, and wool sold at world market prices. The minimum value of protected areas is the opportunity cost of preservation; thus the value is the lower of per-hectare returns to pastureland and cropland, applied to the area under official protection. All benefits were applied over a 25-year time horizon at a 4 percent discount rate.

Summing across all types of rural land, in 1995 the US had a total land value of $1.8 trillion, compared to China's $2.0 trillion. We do not attempt to include the value of urban land. We do not have data to calculate the dynamics of land use change in the period 1995–2000.

e. Results for natural capital. Table 4.1 displays the estimated changes in natural capital, both in quantities and in value terms. In both the US and China, the reductions in oil and natural gas are far greater (in value terms) than those of copper, lead, or phosphate. The increased value of forest offsets about half of the lost value from oil and gas depletion. The reduction in the value of the natural capital stock is about two times larger in the US than in China. However, as a proportion of GDP, the reduction is about five times larger in China.

2   Human capital

The value of the stock of human capital is an important component of a country's wealth. To measure the stock of human capital, we use estimates of average educational attainment contained in an unpublished data set provided by Klenow and Rodríguez-Clare. We will refer to this as the "Klenow and Rodríguez-Clare data."[9]

Table 4.1 Natural capital stocks: Quantities, prices, and values, 1995–2000

United States	Oil	Natural Gas	Copper	Lead	Phosphate	Forests	Land	Total Natural Capital
Capital Stock in 1995	54.91	10,222.03	0.099	0.022	4.200	26.942		
Capital Stock in 2000	40.28	7,495.83	0.090	0.020	4.000	26.976		
Change in Stock 1995–2000	−14.63	−2,726	−0.009	−0.002	−0.200	0.034		
Average Price			2,231	823	42			
Average Extraction Cost			1,513	634	7			
Accounting Price	2.479	0.015	718	189	7	52		
Value of 1995 Stock	136.154	148.694	70.886	4.230	30.829	1,113.923	1,779.705	**3,284.421**
Value of change in Stock	−36.274	−39.656	−6.288	−0.449	−1.465	1.732		**−82.400**

China	Oil	Natural Gas	Copper	Lead	Phosphate	Forests	Land	Total Natural Capital
Capital Stock in 1995	27.88	2,482.17	0.040	0.033	1.315	12.390		
Capital Stock in 2000	22.02	2,366.27	0.037	0.030	1.200	12.450		
Change in Stock 1995–2000	−5.87	−115.90	−0.003	−0.003	−0.115	0.060		
Average Price			2,231	823	42			
Average Extraction Cost			1,717	696	42			
Accounting Price	6.025	0.058		126		30		
Value of 1995 Stock	168.016	144.671		4.192		301.703	2,027.808	**2,646.391**
Value of change in Stock	−35.358	−6.755		−0.398		1.819		**−40.692**

The stock of human capital for an individual, h, is given by

$$h = e^{\rho \cdot A} \tag{12}$$

where ρ is the assumed rate of return on human capital and A is the level of educational attainment. Following Klenow and Rodríguez-Clare (1997), we apply a value of 0.085 for ρ. To find the aggregate stock of human capital, H, we simply multiply h by the population of the county. Rather than use the total population, we exclude children under the age of 17 in the US and children under the age of 11 in China. These age cut-offs are based on the age at which an individual would reach the average education level in each country and are meant to exclude those who have not yet built up their stock of human capital. It is important to point out that the stock of human capital includes the human capital of those not currently in the labor force. Just because an individual is not currently employed does not mean that he or she has no human capital. The measure of the aggregate human capital stock increases both as the average educational attainment increases and as the population over the selected age cut-off increases.

We now need to find the price of a unit of human capital in order to place a value on the stock. Our method is to calculate the rental price for an employed unit of human capital and then to find the average number of working years remaining for the population above the age cut-off. The value of a unit of human capital is the discounted sum of the rental price, r, for the average number of working years remaining.

$$P_{K_H} = \int_{t=0}^{years} r e^{-\rho t} dt \tag{13}$$

The rental price of a unit of human capital is simply the country's total wage bill divided by the employed number of human capital units (not the whole human capital stock). The total wage bill in the US is easily obtained from the national income accounts. China's national income accounting method does not report total wages or compensation, so this is calculated from information provided by the *China Statistical Yearbook*. Employment in both countries is obtained from the Klenow and Rodríguez-Clare data. The average rental price per year for a unit of human capital is $528.11 in China and $12,807.98 in the US.

To calculate the average number of working years remaining for the population over the age cut-off in each country we use data from the

World Health Organization Life Tables and the US Census Bureau IDB demographic data. The calculation depends on the age distribution of the population, the age-specific force of mortality, and the labor market participation rate (probability of employment) at each age. We assume that the force of mortality and the age-specific probability of employment remain constant over time in these calculations. Individuals over the age of 11 in China have on average 21.7 years of work ahead of them, while individuals over the age of 17 in the US have on average 15.7 years of work ahead of them. This gives one unit of human capital a value of $6997 in China and $139,092 in the US.

3   Reproducible capital

The estimated stock of reproducible capital in the US and China are from the Klenow and Rodríguez-Clare data. Our approach to reproducible capital differs from earlier work by the World Bank and by Arrow et al. by accounting for ownership. Some of the stock of reproducible capital in a country is owned by investors outside of that country. Correspondingly, some of the reproducible capital outside a given country is owned by the residents of that country. Our notion of sustainability focuses on the changes in the productive base owned by a given country's residents. Thus it is important to consider changes in a country's net asset position.

In the US, net holdings of international assets are reported by the Bureau of Economic Analysis (BEA). In developing countries, although capital flows are closely monitored, little work has been done on measuring the accumulated stocks of foreign assets and liabilities. We obtain estimates from a recent paper by Philip Lane and Gian Maria Milesi-Ferretti (2007) that constructs net holdings of international assets from balance of payments and other IMF data.

4   Oil capital gains

Our analysis also departs from earlier work in considering capital gains. While capital gains can apply to any capital asset, these gains can be expected to be especially important for stocks of oil. As the world stock of oil decreases, the scarcity rent will increase. Theory suggests an increase at approximately the rate of interest. A country with oil realizes that oil not yet extracted will be worth more tomorrow than today. For each country or region we multiply the stock of oil not extracted during the time period by the difference between price and extraction costs. To calculate the capital gains, we allow the shadow price of oil to increase by 5 percent per year over the period 1995–2000. We apply this increase in the shadow price of oil to the initial (year 1995) oil stock. Thus, the overall change in the value of the oil stock is

$$p_{K_t} I_t + \dot{p}_t K_{t-1} \tag{14}$$

where I_t is the change in the stock from period $t - 1$ to t and $\dot{p}_t$ is the change in the shadow price over this interval.

Capital gains to countries with oil are paid for by countries that consume oil in terms of higher prices. The world total oil capital gains are distributed as a loss to each country in proportion to the fraction of world total oil consumption. The US accounted for 25.7 percent of world oil consumption during the time period and in this calculation we assume that this remains constant over time. Under this assumption, the US pays for 25.7 percent of world total oil capital gains.

5 Environmental capital

The World Bank's Adjusted Net Savings method (World Bank 2006) deducted damages caused by climate change from each national account proportional to that nation's emissions. In other words, the US national account was deducted for the damages caused by the 1.5 billion tons of carbon the US emitted in 2000 (wherever on the planet those damages occurred). The marginal social cost of carbon used in the World Bank method of $25 per metric ton carbon was based on Fankhauser's 1994 article.

Our method changed both the approach and damage estimates. Current models anticipate unequal global distribution of damages from climate change. Therefore, while the US should be morally responsible for compensating other nations for the damage its emissions cause, it is the damage to US assets that should be deducted from its national accounts. We therefore redistribute the global damages based on recent estimates. Furthermore, we actualize the marginal social cost of carbon based on new estimates.

To determine the portion of global damages because of climate change that the US and China will suffer, we utilize Nordhaus and Boyer's (2000) study, which estimates the impacts of various climate change scenarios on economic sectors. We use the most conservative scenario analyzed, corresponding to a doubling of atmospheric concentrations of CO_2-equivalent gases. This scenario is a standard multimodal assessment in the Intergovernmental Panel on Climate Change Third Assessment Report. The physical results of the "greenhouse effect" from this level of pollution are constrained to a mean surface temperature change of 2.5 degrees Celsius over the entire terrestrial environment, where temperature change is latitude-dependent to reflect results of general-circulation models.

Based on this likely (but simplified) scenario, Nordhaus and Boyer apportion the damages to each country as follows: The US will suffer losses of 0.45 percent of its GDP, China 0.22 percent of its GDP, while the world will suffer damages of 1.5 percent of global production. We multiplied each country's expected damage by its GDP in 2000 (from World Development Indicators), and global damage by global GDP in 2000. We then calculated the portion of global damages that each country will suffer. The US will shoulder 9 percent of global loss, and China 1 percent. We use this geographically linked method to determine the portion of global loss each country will suffer, and we now need to calculate the loss because of emissions in the period 1995–2000. To do so, we use updated estimates of the social cost of carbon to calculate global losses.

To calculate global losses because of emissions from 1995–2000, we extracted global carbon emissions data from the World Development Indicators (2005). We converted the data from tons CO_2 to tons C equivalent by multiplying the tons of CO_2 by the ratio of the molecular weight of C to CO_2 (12/44). Using a recent survey by Tol (2005) on the range of marginal damage estimates in the literature, we assigned a conservative marginal social cost of \$50 per ton carbon. This damage estimate is the mean of all peer-reviewed studies analyzed by Tol, and far below the recent estimates by other, more comprehensive studies (see, e.g., Stern 2006). Global emissions of 31 billion tons from the 5-year period of 1996–2000[10] therefore resulted in global damages of \$1612 billion (in year 2000 dollars).

We multiplied the percentage of global loss that each country will suffer (9.32 percent for the US, 0.5 percent for China) by the total global damages calculated above (\$1612 billion) to get the damages suffered by each nation because of its emissions in 1995–2000. As such, the US account was deducted \$150.2 billion and China \$8.1 billion.

6 *Overall changes in capital—comprehensive investment*

Table 4.2 consolidates the changes in all of the forms of capital we have considered. In the US, the increases in human and reproducible capital far outweigh the reductions in natural capital and the net capital losses associated with rising oil prices. Thus, according to this measure, comprehensive investment—the change in the value of the overall capital stock—is positive. For China, comprehensive investment also appears

Table 4.2 Comprehensive investment and its components

United States	Total Natural Capital	Human Capital	Reproducible Capital	Oil Net Capital Gains	Carbon Damages	Sum: Comprehensive Investment
Capital Stock in 1995		0.5419326	13443.5100			
Capital Stock in 2000		0.5844646	16002.9400			
Change in Stock 1995–2000		0.04253	2,559.430			
Accounting Price		139,092.484	1			
Value of 1995 Stock	3,284.421	75,378.750	13,443.510			92,106.682
Value of change in Stock	−82.400	5,915.888	2,559.430	−1,367.580	−150.203	6,875.134
Relative Contribution	−1.20%	86.05%	37.23%	−19.89%	−2.18%	
Percent Change						7.46%
Growth Rate						1.45%

China	Total Natural Capital	Human Capital	Reproducible Capital	Oil Net Capital Gains	Carbon Damages	Sum: Comprehensive Investment
Capital Stock in 1995		1.6228843	4093.4500			
Capital Stock in 2000		1.7951992	6311.0100			
Change in Stock 1995–2000		0.172	2,217.560			
Accounting Price		6,997.466	1			
Value of 1995 Stock	2,646.391	11,356.077	4,093.450			18,095.918
Value of Change in Stock	−40.692	1,205.767	2,217.560	−305.850	−8.127	3,068.658
Relative Contribution	−1.33%	39.29%	72.26%	−9.97%	−0.26%	
Percent Change						16.96%
Growth Rate						3.18%

to be positive. According to the table, the increase in human capital and reproducible capital greatly exceeds the loss from depletion of natural capital.

Table 4.2 also shows the relative contribution of each form of capital (natural, human, reproducible) and of oil capital losses to the overall change in comprehensive investment. In both countries, the relative impact of natural capital depletion is fairly small. Capital losses associated with rising oil prices have a larger impact on comprehensive investment than the depletion of natural capital. The largest impacts are from increases in human and reproducible capital, which overwhelm the negative contributions of the other elements.

It should be emphasized that a key element of these calculations is the shadow or accounting price applied to each type of capital. These indicate the rate at which one form of capital can substitute for another. If the shadow prices for natural capital, in particular, are too low (high), our results will understate (overstate) the lost wealth from depletion in natural resource stocks.

It should also be noted that these calculations do not account for many health-related elements. We discuss this issue further in Section III.

B Accounting for population growth and technological change

We next adjust the changes in comprehensive wealth to account for population growth and technological change. The first column of Table 4.3 reproduces the growth rate of comprehensive investment given in Table 4.2. Column 2 indicates the annual population growth rate of the US and China over the interval 1995–2000. Column 3 subtracts this growth rate from the rate in Column 1 to arrive at the per-capita growth rate of comprehensive wealth.

The next columns adjust for technological change, as measured by the rate of growth of total factor productivity (TFP). Under the assumptions indicated in Section 2, the appropriate adjustment for technological change is obtained by adding the TFP growth rate from the initially obtained growth rate of per-capita comprehensive wealth. Column 5 provides the adjusted rate.

The numbers in column 5 are our ultimate indicators of whether the sustainability criterion is met. According to our calculations, both countries satisfy the criterion, as per-capita comprehensive wealth is growing. (Sensitivity analysis to be offered in next version.) In the US, TFP growth (of 1.48 percent) accounts for about 80 percent of the estimated

Table 4.3 Growth rates of per-capita comprehensive wealth, adjusted for technological change

	(1) Comprehensive Wealth Growth Rate	(2) Population Growth Rate	(3) Per Capita Comprehensive Wealth Growth Rate, Accounting for Population Growth [(1)−(2)]	(4) TFP Growth Rate	(5) Per Capita Comprehensive Wealth Growth Rate, Accounting for TFP growth [(3)+(4)]	(6) Per Capita GDP Growth Rate
US	1.45%	1.10%	0.35%	1.48%	1.83%	4.44%
China	3.18%	0.84%	2.34%	2.71%	5.05%	7.38%

1.86 percent growth rate of comprehensive wealth. China displays even faster growth of comprehensive wealth—a rate of over 5 percent. In the case of China, technological change accounts for about half of this fast growth.

When we initiated this study, we were motivated by a concern about the rapid rate of natural resource depletion in China, as well as the continued and extensive levels of air and water pollution. We sought to gain a better sense as to whether, overall, China's recent economic experience is conducive to higher or lower standards of living for future generations. Although this study is incomplete, it suggests that China's very high rates of investment in reproducible capital and human capital, along with a relatively high rate of technological progress, might well outweigh the costs from natural resource depletion and environmental damage. These interpretations must be very tentative, however. It is important to keep in mind that our results do not capture important environmental and health impacts. Currently, the shadow prices for natural capital do not incorporate beneficial externalities from such capital. Thus, the present analysis could well understate the welfare cost from depletion of such capital. Similarly, the shadow prices for reproducible capital do not yet include the environmental and health impacts from such capital. The bias from this latter omission is not immediately clear. To the extent that newer capital is associated with increased emissions and damages to health, our assessment biases upward the change in comprehensive wealth. On the other hand, to the extent that new capital is associated with improvements in health, the omission biases the wealth change in the opposite direction.

IV   Conclusions

This chapter has presented and applied a framework for determining whether a given nation satisfies a reasonable criterion for sustainability. We define sustainability in terms of the capacity to provide well-being to future generations. The principal indicator of this capacity is a comprehensive measure of wealth—one that includes both marketed and nonmarketed assets. The sustainability criterion is satisfied if this comprehensive measure of wealth is increasing on a per-capita basis.

Our framework follows Arrow et al. (2004) in integrating population growth and technological change in the analysis of comprehensive wealth. It offers further methodological improvements by accounting for capital gains, providing a closer assessment of changes in human capital, and addressing potential damages from climate change.

Our initial application of this framework to China and the US suggests that both nations are meeting the sustainability criterion. In the US, increases in human capital and (to a lesser extent) reproducible capital significantly outweigh the adverse wealth effects from natural resource depletion and higher oil prices. In China, investments in reproducible capital contribute the most to increases in genuine wealth, although increases in human capital and (predicted) technological progress also play a significant role. Importantly, China's depletion of natural resources, though very significant, do not have nearly as large an impact on wealth as do the contributions from investments in reproducible and human capital.

These results must be viewed as preliminary and tentative. We have not yet incorporated many important health impacts, which could significantly change the picture. Between 1995 and 2000 life expectancy at birth for the population as a whole increased by 1.6 years in China and 1.2 years in the US. In China the gain was in large measure a reflection of reductions in the under-5 mortality rate (from 46 to 40 deaths per 1000 births), while in the US the major factor would appear to have been reductions in mortality caused by cardiovascular disease. A commonly accepted method for valuing reductions in mortality rates in terms of income is to estimate differences in wages that can be attributed to differences in the risk of death in various occupations. Measured thus, the gains would appear to be very large. For example, Nordhaus (2002) has estimated that during the last 100 years the economic gains in the US from increases in life expectancy were comparable to the growth in nonhealth consumption goods and services. We conjecture from that work that the contribution of improvements in health to the accumulation of comprehensive wealth could be substantial. In future work we intend to estimate that contribution by appealing to a range of approaches to the value of improvements in the health and longevity.

Although ignoring improvements in health biases downward our estimated increases in comprehensive wealth in China and the US, our neglect of a wide range of losses caused by environmental degradation (e.g., soil loss, water stress, increases in atmospheric pollutants) implies the opposite bias.[11] In future work we hope to take account of changes in a wider range of natural capital stocks.

The estimates we have offered in Table 4.3 are marred also by the considerable uncertainties that surround the values of the shadow prices employed here, which determine the rates of convertibility across types of capital. Large uncertainties surround technological change as well.

Despite these limitations, we hope that our efforts will promote more focused thinking about sustainability and its measurement, as well as change people's priors about whether the criterion is being satisfied in the US and China.

Acknowledgment

We are very grateful to Kirk Hamilton and Giovanni Ruta of the World Bank and Peter Klenow of Stanford for very helpful advice and for generously supplying important data for this project. We also thank Kenneth Chay and T. N. Srinivasan for very helpful comments.

Appendix

1   General data

Population

We extracted population from the World Development Indicators (http://devdata.worldbank.org).

GDP

GDP data were obtained from the World Development Indicators. Data are in current US dollars.

TFP

Total Factor Productivity data were obtained from Peter Klenow (unpublished data). These data were employed in Klenow and Rodríguez-Clare (2005) at http://www.klenow.com.

2   Data on natural resources

Oil

The year-end 2004 proved reserves for the US and China were obtained from the *BP Statistical Review of World Energy* (2005). We also obtained oil production (extraction) for both countries in each year from 1995 to 2004 from this publication. The stock of oil in years 1995–2003 is given by adding the production from the previous year to the stock from the previous year:

$$Stock_{t-1} = Stock_t + Production_t$$

The calculated stock values, along with production and consumption for each country, are given in Table A.4.4.

Table A.4.1 Population in US and China, 1995–2000

	1995	1996	1997	1998	1999	2000
US	266,278,000	269,393,984	272,656,992	275,854,016	279,040,000	282,224,000
China	1,204,855,040	1,217,549,952	1,230,075,008	1,241,934,976	1,253,735,040	1,262,644,992

Table A.4.2 Gross domestic product, US and China, 1995–2000

	1995	1996	1997	1998	1999	2000
US	7,342,300,069,888	7,762,299,846,656	8,250,900,086,784	8,694,599,778,304	9,216,199,753,728	9,764,800,036,864
China	700,277,784,576	816,489,824,256	898,243,690,496	946,300,846,080	991,355,666,432	1,080,741,396,480

Table A.4.3 TFP for US and China, 1995–2000

	1995	1996	1997	1998	1999	2000
US	620.99	628.15	639.57	647.97	656.91	668.26
China	187.59	190.92	193.71	198.22	200.98	214.38

Table A.4.4 Oil stock, production, and consumption in billions of barrels 1995–2004

US	1995	1996	1997	1998	1999	2000	2001	2002	2003	2004
Stock	54.91	51.88	48.86	45.94	43.11	40.28	37.48	34.70	32.00	29.35
Production	3.04	3.04	3.02	2.92	2.82	2.83	2.80	2.78	2.70	2.65
Consumption	6.47	6.70	6.80	6.90	7.12	7.21	7.17	7.21	7.31	7.51
China	1995	1996	1997	1998	1999	2000	2001	2002	2003	2004
Stock	27.88	26.72	25.55	24.38	23.21	22.02	20.81	19.59	18.35	17.07
Production	1.09	1.16	1.17	1.17	1.17	1.19	1.21	1.22	1.24	1.28
Consumption	1.24	1.34	1.44	1.48	1.61	1.82	1.84	1.96	2.11	2.45

Source: *BP Statistical Review of World Energy* and authors' calculations.

The wealth of a country includes the value of the stock of oil net of extraction costs. If the price of a barrel of oil is $30, but it costs $20 for each barrel that is extracted, the value of a barrel of oil to the country is $10. This type of calculation requires some simplifying assumptions. First, oil is not a homogenous good. There are, in fact, many different grades of oil with corresponding prices. Second, over the time period in this study, 1995–2000, the price of any particular grade of oil varies significantly. We average both over oil grades and over time to calculate an average price of oil for the 1995–2000 period. To calculate this price, we use the real average of spot prices over 1995 to 2000 for four types of crude: Dubai, Brent, Nigerian Forcados, and West Texas Intermediate. We adjust the prices by the Consumer Price Index for urban consumer CPI-U to account for inflation before averaging over time. The prices are reported in Table A.4.5.

Table A.4.5 Spot prices for crude oil and average world price for 1995–2000

	1995	1996	1997	1998	1999	2000
Dubai	16.10	18.52	18.23	12.21	17.25	26.20
Brent	17.02	20.67	19.09	12.72	17.97	28.50
Nigerian Forcados	17.26	21.16	19.33	12.62	18.00	28.42
West Texas	18.42	22.16	20.61	14.39	19.31	30.37
Average Price	17.20	20.63	19.32	12.99	18.13	28.37
Average Real Price	18.67	21.98	20.24	13.46	18.53	28.37

Source: *BP Statistical Review of World Energy* (2005).

Using this method, the average world price of oil for 1995–2000 is $20.21 per barrel.

The estimated cost of extraction is obtained from the World Bank (2006). The World Bank uses several studies of the costs of oil extraction and combines them into a large database. While there are several estimates of the costs of oil extraction in the US, there is none for China. Because there is no estimate, we assume that the cost of oil extraction in China is 80 percent of the cost in the US. Oil production in China is actually quite similar to US oil production. China has a large number of offshore facilities and even though labor costs are considerably less, the oil industry is particularly capital intensive. The 80 percent assumption is based on conversations with energy experts, but even so, it is a fairly arbitrary assumption. Our estimate of the cost of oil extraction is $17.73 per barrel in the US and $14.18 per barrel in China. The calculated rent from oil is $2.48 in the US and $6.03 in China.

Natural gas.   The end of year 2004 proved reserves for the US and China were obtained from the *BP Statistical Review of World Energy* (2005). We also obtained natural gas production (extraction) for both countries in each year from 1995 to 2004 from this publication. The stock of natural gas in years 1995–2003 is given by adding the production from the previous year to the stock from the previous year:

$$Stock_{t-1} = Stock_t + Production_t$$

The calculated stock values, along with production and consumption for each country, are given in Table A.4.6.

The wealth of a country includes the value of the stock of natural gas net of extraction costs. We average both over natural gas source and over time to calculate an average price of natural gas for the 1995–2000 period. To calculate this price, we use the real average of spot prices over 1995 to 2000 for four natural gas sources. We adjust the prices by the CPI-U to account for inflation before averaging over time. The prices are reported in Table A.4.7. Using this method, the average world price of natural gas for 1995–2000 is $102.42 per thousand cubic meters.

The estimated cost of natural gas production is obtained from the World Bank (2005). The World Bank uses several studies of the costs of oil extraction and combines them into a large database. While there are several estimates of the costs of natural gas extraction in

Table A.4.6   Natural gas stock, production, and consumption in billion cubic meters 1995–2004

US	1995	1996	1997	1998	1999	2000	2001	2002	2003	2004
Stock	10218.8	9677.1	9134.0	8584.8	8043.2	7492.6	6926.8	6382.5	5832.9	5290.0
Production	534.3	541.7	543.1	549.2	541.6	550.60	565.8	544.3	549.6	542.9
Consumption	638.0	649.6	653.2	642.2	644.3	669.70	641.4	661.6	645.3	646.7
China	1995	1996	1997	1998	1999	2000	2001	2002	2003	2004
Stock	2483.3	2463.4	2441.2	2418.9	2394.6	2367.4	2337.1	2305.2	2270.8	2230.0
Production	17.6	19.9	22.2	22.3	24.3	27.2	30.3	31.9	34.4	40.8
Consumption	17.7	17.7	19.3	19.3	21.4	24.5	27.8	29.6	32.8	39.0

Source: *BP Statistical Review of World Energy* and authors' calculations.

Table A.4.7   Spot prices for natural gas in thousand cubic meters and average world price, 1995–2000

	1995	1996	1997	1998	1999	2000
US	61.53	100.49	92.12	75.73	82.65	154.01
UK	–	67.36	73.91	69.91	59.71	97.58
Japan	125.98	133.26	142.36	111.05	114.32	171.85
European Union	86.29	88.47	96.48	82.28	65.54	118.33
Average Price	**91.27**	**97.39**	**101.22**	**84.74**	**80.56**	**135.44**
Average Real Price	**99.09**	**103.77**	**106.08**	**87.84**	**82.31**	**135.44**

Source: *BP Statistical Review of World Energy* (2005).

the US, again, there is none for China. Because there is no estimate, we assume that the cost of natural gas extraction in China is equal to the average cost of natural gas extraction in the world. Our estimate of the cost of natural gas production is $87.88 per thousand cubic meters in the US and $44.14 per thousand cubic meters in China. The calculated rent from natural gas is $14.55 in the US and $58.28 in China.

Metals and minerals. To get stock, production, cost, and price data, we used a number of sources and applied estimation techniques across all metals and minerals. In general, we used the same data and methods described in the World Bank's *Manual for Calculating Net Adjusted Savings* (Bolt et al. 2002) and *Where Is the Wealth of Nations* (World Bank 2006). We briefly describe these below.

We set the stock, or the reserves, of all metals and minerals in 2000 equal to the reserve base, that is, the proven reserve plus the probable reserve, as reported in the US Bureau of Mines' Mineral Commodity Summaries. Proven reserves are profitably exploitable under current economic conditions, while probable reserves are less certain, but also thought to be exploitable under current economic conditions at some point in the future.

Production numbers for most minerals are fairly complete in the World Bank dataset. These are based on USGS numbers published in their *Mineral Commodities Summary* and/or *Minerals Yearbook*, extrapolated linearly to fill in gaps of missing years.

Given the estimated stock of an exhaustible resource at the end of year t, the stock at the end of year $t-1$ is given by

$$Stock_{t-1} = Stock_t + Production_t$$

where the production in year t measures the amount of the resource extracted between year $t-1$ and year t.

Extraction cost data are proprietary and therefore very difficult to obtain. We used the World Bank dataset, which compiled data on a wide array of sources and expert opinion.

World market prices came from the United Nations Conference on Trade and Development (UNCTAD), *Monthly Commodity Price Bulletin*.

If, for all years in 1995–2000, price was greater than cost, we averaged the difference between annual real price and annual real extraction cost, adjusted to year 2000 dollars, to obtain our measure of the average unit rent for the period 1995–2000.

Copper

Table A.4.8a US copper stock, production, and extraction costs

US		1995	1996	1997	1998	1999	2000
Stock	metric tons	98,760,000	96,840,000	94,900,000	93,040,000	91,440,000	90,000,000
Production	metric tons	1,850,000	1,920,000	1,940,000	1,860,000	1,600,000	1,440,000
Extraction Costs	*US$/metric ton*	*1,394*	*1,420*	*1,444*	*1,460*	*1,481*	*1,513*
Extraction costs	*Y2000 $/metric ton*	*1,513*	*1,513*	*1,513*	*1,513*	*1,513*	*1,513*
						average	**1,513**

Table A.4.8b China copper stock, production, and extraction costs

China		1995	1996	1997	1998	1999	2000
Stock	metric tons	39,531,100	39,092,000	38,596,000	38,110,000	37,590,000	37,000,000
Production	metric tons	445,200	439,100	496,000	486,000	520,000	590,000
Extraction Costs	*US$/metric ton*	*1,581*	*1,611*	*1,638*	*1,656*	*1,680*	*1,717*
Extraction costs	*Y2000 $/metric ton*	*1,717*	*1,717*	*1,717*	*1,717*	*1,717*	*1,717*
						average	**1,717**

Table A.4.8c   World market price for copper, 1995–2000

		1995	1996	1997	1998	1999	2000
Average Price	US$/metric ton	2,981	2,355	2,294	1,671	1,607	1,862
Price Y2000	Y2000 $/metric ton	3,236	2,509	2,404	1,732	1,642	1,862
						average	2,231

Table A.4.8d   Average unit rent for copper, 1995–2000

	1995	1996	1997	1998	1999	2000	average
US	1,723	996	891	219	129	349	718
China	1,519	793	687	15	0	145	..

Lead

Table A.4.9a US lead stock, production, and extraction costs

US		1995	1996	1997	1998	1999	2000
Stock	metric tons	22,373,000	21,937,000	21,478,000	20,985,000	20,465,000	20,000,000
Production	metric tons	394,000	436,000	459,000	493,000	520,000	465,000
Extraction Costs	*US$/metric ton*	*584*	*595*	*605*	*611*	*620*	*634*
Extraction costs	*Y2000 $/metric ton*	*634*	*634*	*634*	*634*	*634*	*634*
						average	**634**

Table A.4.9b China lead stock, production, and extraction costs

China		1995	1996	1997	1998	1999	2000
Stock	metric tons	33,144,000	32,501,000	31,789,000	31,209,000	30,660,000	30,000,000
Production	metric tons	520,000	643,000	712,000	580,000	549,000	660,000
Extraction Costs	*US$/metric ton*	*641*	*654*	*665*	*672*	*682*	*696*
Extraction costs	*Y2000 $/metric ton*	*696*	*696*	*696*	*696*	*696*	*696*
						average	**696**

Table A.4.9c   World market price for lead, 1995–2000

		1995	1996	1997	1998	1999	2000
Average Price	US$/metric ton	778	922	825	763	733	707
Price Y2000	Y2000 $/metric ton	844	983	864	791	749	707
						average	823

Table A.4.9d   Average unit rent for lead, 1995–2000

	1995	1996	1997	1998	1999	2000	average
US	210	349	231	157	115	73	189
China	148	286	168	94	52	10	126

Phosphate

Table A.4.10a   US phosphate stock, production, and extraction costs

US		1995	1996	1997	1998	1999	2000
Stock	metric tons	4,199,600,000	4,154,200,000	4,108,300,000	4,072,200,000	4,036,100,000	4,000,000,000
Production	metric tons	43,500,000	45,400,000	45,900,000	36,100,000	36,100,000	36,100,000
Extraction Costs	*US$/metric ton*	*32*	*33*	*33*	*34*	*34*	*35*
Extraction costs	*Y2000 $/metric ton*	*35*	*35*	*35*	*35*	*35*	*35*
						average	35

Table A.4.10b   China phosphate stock, production, and extraction costs

China		1995	1996	1997	1998	1999	2000
Stock	metric tons	1,314,500,000	1,293,500,000	1,269,000,000	1,246,000,000	1,223,000,000	1,200,000,000
Production	metric tons	19,300,000	21,000,000	24,500,000	23,000,000	23,000,000	23,000,000
Extraction Costs	*US$/metric ton*	*46*	*47*	*48*	*48*	*49*	*50*
Extraction costs	*Y2000 $/metric ton*	*50*	*50*	*50*	*50*	*50*	*50*
						average	50

Table A.4.10c   World market price for phosphate, 1995–2000

		1995	1996	1997	1998	1999	2000
Average Price	US$/metric ton	35	38	41	42	44	44
Price Y2000	Y2000 $/metric ton	38	40	43	44	45	44
						average	42

Table A.4.10d   Average unit rent for phosphate, 1995–2000

	1995	1996	1997	1998	1999	2000	average
US	3	6	8	9	10	9	7
China	0	0	0	0	0	0	..

Forests.   FAO calculates each country's total forest stock from estimates of average stock per hectare for each region applied to the total forested area in each nation. FRA 2005 data confirm that the productive functions of global forest resources have not changed significantly in the past 15 years; the density of wood per hectare and total growing stock are relatively steady at the global level.

We obtained total cubic meters of commercially available forests from the Food and Agriculture Organization (FAO) Forestry Resources Assessment (FRA) (FAO 2005). These data include volumes of growing stock of forests and other wooded resources in 1990, 2000, and 2005, and designate the amount of total stock that is commercially available. FAO calculates each country's total forest stock from estimates of average stock per hectare for each region applied to the total forested area in each nation.

We calculated a linear growth rate between 1990 and 2000 to get stock data for individual years. The difference in stock from one year to the next is assumed to be "produced" (if negative) or "afforested" (if positive). For both the US and China, stock increased over 1995–2000. For 2000, the production was set equal to production in 1999 (as no stock was calculated for 2001).

$$Production_t = Stock_t - Stock_{t+1}$$

Rental price for forests equals the weighted average market price of the types of wood minus the extraction and afforestation costs. World

Table A.4.11a US commercially available forest stock, 1995–2000

US		1995	1996	1997	1998	1999	2000
Stock	billion cubic meters	26.9425	26.9492	26.9559	26.9626	26.9693	26.9760
Production	billion cubic meters	0.0067	0.0067	0.0067	0.0067	0.0067	0.0067

Table A.4.11b China commercially available forest stock, 1995–2000

China		1995	1996	1997	1998	1999	2000
Stock	billion cubic meters	12.3902	12.4022	12.4141	12.4261	12.4380	12.4500
Production	billion cubic meters	0.0120	0.0120	0.0120	0.0120	0.0120	0.1395

Table A.4.11c   Weighted average roundwood prices, US and China, 1995–2000

		1995	1996	1997	1998	1999	2000
USA	current US$/cum	147	149	135	94	106	109
	Y2000 $/cum	160	159	141	98	109	109
						average	129
China	current US$/cum	65	67	60	53	50	55
	Y2000 $/cum	71	71	62	55	51	55
						average	61

Table A.4.11d   Average unit rent for forests, 1995–2000, Y2000 dollars

	1995	1996	1997	1998	1999	2000	average
US	64	63	56	39	43	44	52
China	35	35	31	27	25	28	30

market prices are a weighted average for fuel and roundwood, according to the formula:

$$P_r = Q_f{}^*(P_f) + (1 - Q_f)^*(P_e)$$

where,

P_r = Weighted average price of roundwood
P_f = Price of fuelwood
P_e = Export price of industrial roundwood (which does not reflect fuelwood)
Q_f = Fuelwood quotient, i.e., percentage of total roundwood production that is fuelwood

With these inputs, we can calculate the average unit rent for forests.

Land. *Where Is the Wealth of Nations* (2005) provides the following estimates of land values.

Table A.4.11e   Components of land value, $ Per Capita, 2000

	Nontimber forest resources	Protected Areas	Cropland	Pasture land	Total Land Value
US	238	1,651	2,752	1,665	6,306
China	29	27	1,404	146	1,606

Carbon damages

Table A.4.11f Carbon emissions, 000 tons Carbon

	1995	1996	1997	1998	1999	2000	Total (96–00)
US	1,416,796	1,438,965	1,485,985	1,499,690	1,501,796	1,527,684	7,454,120
China	872,086	911,655	898,166	850,076	770,461	761,032	4,191,390
Global	6,053,892	6,053,892	6,053,892	6,053,892	6,053,892	6,053,892	30,269,462

Oil capital gains.   In the oil section, we try to measure the change in wealth because of changes in the quantity of oil given a fixed price. Here, we try to measure the change in wealth because of changes in price given a fixed quantity. A country with a stock of oil chooses how much to extract each year, subject to a capacity constraint. As the world stock of oil decreases, the price increases driving up the scarcity rent.

We assume that the scarcity rent increases by 5 percent per year. Oil that was not extracted during the 1995–2000 period then would have increased in value by 27.6 percent. For each country (or region where the data does not allow desegregation) we calculate the value of the stock of oil remaining in 2000 and then multiply this by the assumed increase in prices to find the capital gains.

Capital gains to countries with oil are paid for by countries that consume oil in terms of higher prices. The world total oil capital gains are distributed as a loss to each country in proportion to the fraction of world total oil consumption. Since the US accounted for 25.7 percent of world oil consumption, it is assigned a loss equal to 25.7 percent of world total oil capital gains because of the increasing price of oil.

We obtain the level of oil reserves in 2004 from the *BP Statistical Review of World Energy* (2005) and then calculate the level of oil reserves in 2000 in the same method as described in the oil section above. The production data used to make these calculations and the consumption data used in calculating the percent of world consumption is also obtained from the *BP Statistical Review of World Energy* (2005). The extraction cost data is from the World Bank (2005). Extraction cost data is missing for some countries.

3   Human capital

We use an estimate of the average educational attainment reported in Klenow and Rodríguez-Clare (2005) to construct a measure the stock of human capital. The stock of human capital for an individual, h, is given by

$$h = e^{0.085 \cdot \text{(educational attainment)}}$$

where 0.085 is the assumed rate of return on human capital. To find the aggregate stock of human capital, H, we simply multiply h by the population of the county. Rather than use the total population, we exclude children under the age of 11 in China and under the age of

Table A.4.12   Oil reserves, consumption, and capital gains, 1995–2000

Country or Region	Extraction Cost 95–00 Ave	Rent per Barrel 95–00 Ave	2000 Reserves Billions Barrels	Gross Capital Gain $ Billions	Consumption % of World (95–00)	Net Capital Gain $ Billions
US	17.7	2.51	40.28	$27.93	25.73%	−$1,367.38
Canada	24.4	0.00	20.94	$0.00	2.63%	−$142.50
Mexico	4.3	15.91	20.06	$88.16	2.43%	−$43.37
Argentina	15.3	4.91	3.82	$5.18	0.56%	−$25.04
Brazil	17.7	2.51	13.36	$9.26	2.37%	−$119.05
Colombia	21.7	0.00	2.38	$0.00	0.33%	−$17.72
Ecuador	4.2	16.01	5.69	$25.18	0.18%	$15.62
Peru	10.7	9.51	1.07	$2.80	0.20%	−$8.18
Trinidad & Tobago	8.1	12.11	1.21	$4.04	0.01%	$3.50
Venezuela	4.3	15.91	81.52	$358.32	0.67%	$322.04
Other S. & C. America	18.0	2.21	1.60	$0.98	1.55%	−$82.97
Azerbaijan	10.0	10.21	7.44	$20.99	0.16%	$12.47
Denmark	15.0	5.21	1.86	$2.67	0.28%	−$12.73
Italy	15.0	5.21	0.88	$1.27	2.60%	−$139.96
Kazakhstan	10.0	10.21	41.14	$116.04	0.25%	$102.66
Norway	15.0	5.21	14.37	$20.69	0.28%	$5.24
Romania	10.0	10.21	0.65	$1.83	0.31%	−$14.85
Russia	8.1	12.11	83.88	$280.63	3.42%	$95.04
Turkmenistan	10.0	10.21	0.81	$2.29	0.10%	−$2.96
United Kingdom	17.4	2.81	7.77	$6.03	2.31%	−$119.47
Uzbekistan	10.0	10.21	0.83	$2.34	0.18%	−$7.49
Other Europe & Eurasia	15.0	5.21	2.60	$3.74	15.53%	−$838.27

Iran	0.8	19.41	137.88	$739.40	1.73%	$645.43
Iraq	0.8	19.41	117.77	$631.57	0.38%	$610.96
Kuwait	1.6	18.61	102.07	$524.79	0.26%	$510.95
Oman	4.3	15.91	6.81	$29.92	0.07%	$26.12
Qatar	4.0	16.21	16.47	$73.76	0.07%	$69.86
Saudi Arabia	0.8	19.41	276.63	$1,483.45	1.95%	$1,377.64
Syria	4.0	16.21	3.95	$17.71	1.00%	−$36.53
United Arab Emirates	6.0	14.21	101.28	$397.61	0.40%	$375.97
Yemen	6.0	14.21	3.50	$13.73	0.40%	−$7.97
Other Middle East			0.00	$0.00	0.50%	−$27.12
Algeria	15.0	5.21	14.30	$20.59	0.27%	$5.82
Angola	15.0	5.21	10.06	$14.47	0.05%	$11.76
Chad	15.0	5.21	0.97	$1.39	0.01%	$0.85
Rep. of Congo	15.0	5.21	2.14	$3.09	0.01%	$2.54
Egypt	11.9	8.31	4.62	$10.61	0.72%	−$28.48
Equatorial Guinea	15.0	5.21	1.64	$2.36	0.01%	$1.82
Gabon	13.0	7.21	2.67	$5.31	0.02%	$4.23
Libya	4.3	15.91	41.22	$181.21	0.29%	$165.48
Nigeria	4.3	15.91	38.51	$169.28	0.37%	$149.22
Sudan	15.0	5.21	6.67	$9.60	0.08%	$5.26
Tunisia	15.0	5.21	0.74	$1.07	0.11%	−$4.90
Other Africa	15.0	5.21	0.80	$1.15	0.89%	−$47.12

Table A.4.12   (Continued)

Country or Region	Extraction Cost 95–00 Ave	Rent per Barrel 95–00 Ave	2000 Reserves Billions Barrels	Gross Capital Gain $ Billions	Consumption % of World (95–00)	Net Capital Gain $ Billions
Australia	15.0	5.21	4.98	$7.17	1.11%	−$52.93
Brunei	15.0	5.21	1.35	$1.94	0.02%	$0.86
China	14.16	6.05	22.02	$36.81	6.32%	−$305.80
India	15.0	5.21	6.70	$9.65	2.82%	−$143.03
Indonesia	7.1	13.11	6.50	$23.53	1.35%	−$49.57
Malaysia	4.3	15.91	5.51	$24.23	0.59%	−$7.80
Thailand	15.0	5.21	0.79	$1.13	1.03%	−$54.50
Vietnam	15.0	5.21	3.49	$5.03	0.26%	−$9.07
Other Asia Pacific	15.0	5.21	0.90	$1.30	14.86%	−$804.60
			World Total	$5,423.23	100.00%	$0.00

Source: PB Review (2005) and estimates of the country-specific average extraction costs (if unavailable a value of 15 is used for the calculations).

17 in the US because they have not yet built up their stock of human capital.

The average educational attainment increased during the 1995–2000 period. China experienced a growth of 4 percent, while the US had slightly more than 1 percent growth during the period (see Table A.4.13). The age 10 or more population in each country also increases over the 1995–2000 time period. China experienced 8.6 percent growth and the US experienced 6.7 percent growth (see Table A.4.13). Note that our measure of the aggregate human capital stock increases both as the average educational attainment increases and as the population of age 10 or more increases.

As Table A.4.13 shows, the aggregate stock of human capital increased almost 11 percent in China and slightly more than 8 percent in the US.

We now need to find the price of a unit of human capital in order to place a value on the stock. The methodology here is to first calculate the rental price for an employed unit of human capital and then to find the average number of working years remaining for the population age 10 or more. The value of a unit of human capital is the discounted sum of the rental price, r, for the average number of working years remaining.

$$P_{K_H} = \int_{t=0}^{years} re^{-\rho t} dt$$

The rental price of a unit of human capital is simply the country's total wage bill divided by the employed number of human capital units (not the whole human capital stock). The total wage bill in the US is easily obtained from the national income accounts. China's national income accounting method does not report total wages or compensation, so this is calculated from information provided by the *China Statistical Yearbook*. Employment in both countries is obtained from Klenow and Rodríguez-Clare (2005). We use the average price for the time period. The average rental price per year for a unit of human capital is $528.11 in China and $12,807.98 in the US.

4   Reproducible capital

Table A.4.20 shows the measure of the stock of reproducible capital in the US and China that we obtained from Klenow and Rodríguez-Clare (2005).

Table A.4.13   Education, population, and human capital stock, 1995–2000

Year	China				US			
	Average Education Attainment	Average Human Capital (h)	Population Age 11 + (thousands)	Human Capital Stock (H)	Average Education Attainment	Average Human Capital (h)	Population Age 17 + (thousands)	Human Capital Stock (H)
1995	6.111	1.68108	965,382	1,622,884,318	11.892	2.74785	197,221	541,932,590
1996	6.160	1.68809	979,216	1,653,004,350	11.923	2.75510	199,569	549,831,670
1997	6.209	1.69514	997,315	1,690,588,728	11.955	2.76261	202,139	558,431,592
1998	6.257	1.70207	1,013,298	1,724,703,967	11.986	2.76990	204,683	566,951,779
1999	6.306	1.70917	1,029,251	1,759,164,853	12.018	2.77744	207,250	575,623,498
2000	6.355	1.71631	1,045,964	1,795,199,173	12.049	2.78477	209,879	584,464,633

Source: Population: US Census Bureau International Database (http://www.census.gov/ipc/www/idb/).

Table A.4.14 Total wage bill for China and the US 1995–2000

Year	China Avg. Wage ($US)	China Employment (thousands)	China Wage Bill ($billions)	US Wage Bill ($billions)
1995	684	727,832	497.8	4,177
1996	772	735,241	567.6	4,387
1997	805	742,369	597.6	4,665
1998	930	749,197	696.8	5,020
1999	1,038	756,055	784.8	5,352
2000	1,166	763,855	886.1	5,783

Note: US Wage Bill = Total compensation from national income account, BEA.
Average Wage (nominal) Source: *China Statistical Yearbook* 2002: Table 5–20 (http://www. stats.gov.cn/english/statisticaldata/yearlydata/YB2002e/ml/indexE.htm).
Employment Source: Klenow, P. and Rodríguez-Clare, A. Handbook Chapter (Data Appendix).

This is a measure of the reproducible capital located in the country. However, because of international investment, some of the stock of reproducible capital in a country is owned by investors outside of that country. In recent years, the US has experienced a large trade deficit while China has experienced a trade surplus (see Table A.4.21). This implies that other countries have a claim on a large stock of the reproducible capital in the US, while China is in the opposite position. The portion of the stock of reproducible capital that is not owned by the country in which it is located cannot be counted as part of its wealth. It is the stock of reproducible capital owned by a country regardless of its physical location that is a component of wealth.

In the US, net holdings of international assets are reported by the BEA. In developing countries, although capital flows are closely monitored, little work has been done on measuring the accumulated stocks of foreign assets and liabilities. We obtain estimates from a recent paper by Philip Lane and Gian Maria Milesi-Ferretti (2007) that constructs net holdings of international assets from balance of payments and other IMF data.

Table A.4.15   Rental price of human capital in China and the US, 1995–2000

Year	China			US		
	Wage Bill ($billions)	Human Capital Stock (H)	Rental price of one unit of human capital	Wage Bill ($billions)	Human Capital Stock (H)	Rental price of one unit of human capital
1995	497.84	1,223,541,238	406.88	4177	366,310,386	11,402.90
1996	567.61	1,241,156,279	457.32	4387	372,957,055	11,762.75
1997	597.61	1,258,419,442	474.89	4665	379,093,145	12,305.68
1998	696.75	1,275,185,340	546.39	5020	384,372,859	13,060.24
1999	784.79	1,292,229,879	607.31	5352	390,452,142	13,707.19
2000	886.07	1,311,009,179	675.87	5783	395,848,675	14,609.12

Notes: China: Average rental price of one unit of human capital = **$528.11**.
US: Average rental price of one unit of human capital = **$12,807.98**.

Table A.4.16 Population, mortality, and years of work remaining for males in China

	Population	Active Percentage	Active Population	Mortality Probability	Avg. Work Remaining
0–4	51091760	0	0	0.03759	43.0
5–9	54254180	0	0	0.00314	44.7
10–14	62010528	0	0	0.00269	44.8
15–19	52665280	63.86	33632048	0.00578	41.7
20–24	50961980	92.8	47292717	0.00718	37.3
25–29	62313660	98.61	61447500	0.00674	32.7
30–34	64799528	99.02	64164493	0.00732	28.0
35–39	53963232	99.15	53504545	0.00946	23.2
40–44	43936400	98.95	43475068	0.01406	18.5
45–49	44348360	97.94	43434784	0.02256	13.8
50–54	32585930	93.55	30484138	0.0368	9.5
55–59	24418930	83.88	20482598	0.05939	5.7
60–64	21478200	63.75	13692353	0.09408	2.8
65–69	17625390	33.59	5920369	0.14642	1.4
70–74	11991350	33.59	4027894	0.22671	0.0
75–79	6863370	0	0	0.34238	0.0
80–84	3015910	0	0	0.49999	0.0
85–89	903120	0	0	0.67919	0.0
90–94	163110	0	0	0.79408	0.0
95–99	16994	0	0	0.85642	0.0
100+	1091	0	0	1	0.0

Sources: World Health Organization Life Tables 2000; US Census Bureau IDB demographic data 1990.

Table A.4.17 Population, mortality, and years of work remaining for females in China

	Population	Active Percentage	Active Population	Mortality Probability	Avg. Work Remaining
0–4	45892300	0	0	0.04389	35.1
5–9	48638800	0	0	0.00264	36.7
10–14	56168520	0	0	0.0018	36.8
15–19	48594528	71.4	34696493	0.00238	33.3
20–24	47770000	91.68	43795536	0.0032	28.8
25–29	59211928	91.56	54214441	0.0041	24.3
30–34	62217712	91.3	56804771	0.00529	19.9
35–39	51377952	91.28	46897795	0.00743	15.4
40–44	40608800	88.37	35885997	0.01049	11.1
45–49	41938880	81.12	34020819	0.01603	7.2
50–54	30147800	62	18691636	0.02465	4.2

138

Table A.4.17 (Continued)

	Population	Active Percentage	Active Population	Mortality Probability	Avg. Work Remaining
55–59	22592370	45.07	10182381	0.03787	2.0
60–64	20259450	27.44	5559193	0.06183	0.8
65–69	17690780	8.44	1493102	0.10341	0.4
70–74	13294740	8.44	1122076	0.17766	0.0
75–79	9051270	0	0	0.29618	0.0
80–84	4898760	0	0	0.45765	0.0
85–89	1955500	0	0	0.64713	0.0
90–94	591650	0	0	0.77727	0.0
95–99	112775	0	0	0.84928	0.0
100+	11490	0	0	1	0.0

Sources: World Health Organization Life Tables 2000; US Census Bureau IDB demographic data 1990.

Table A.4.18 Population, mortality, and years of work remaining for males in the US

	Population	Active Percentage	Active Population	Mortality Probability	Avg. Work Remaining
0–4	10265130	0	0	0.00936	40.3
5–9	10716110	0	0	0.00093	40.6
10–14	10508530	0	0	0.00123	40.7
15–19	10125050	53.2	5386527	0.00474	38.1
20–24	9445860	82.5	7792835	0.00706	34.1
25–29	9523560	92.94	8851197	0.00706	29.7
30–34	10369200	93.44	9688980	0.00784	25.3
35–39	11693900	92.74	10844923	0.01043	20.8
40–44	11610640	91.96	10677145	0.01499	16.4
45–49	10217120	90.76	9273058	0.02274	12.2
50–54	8788100	86.9	7636859	0.03233	8.1
55–59	6703200	77.87	5219782	0.04915	4.5
60–64	5236380	54.28	2842307	0.07541	2.0
65–69	4431380	27.53	1219959	0.11333	0.8
70–74	3927820	17.3	679513	0.16984	0.0
75–79	3059800	7.3	223365	0.24554	0.0
80–84	1835290	0	0	0.36599	0.0
85–89	836240	0	0	0.52072	0.0
90–94	283400	0	0	0.65771	0.0
95–99	68020	0	0	0.76489	0.0
100+	10130	0	0	1	0.0

Sources: World Health Organization Life Tables 2000; US Census Bureau IDB demographic data 1990.

Table A.4.19 Population, mortality, and years of work remaining for females in the US

	Population	Active Percentage	Active Population	Mortality Probability	Avg. Work Remaining
0–4	9777090	0	0	0.00766	33.9
5–9	10215890	0	0	0.00072	34.1
10–14	10024470	0	0	0.00082	34.2
15–19	9661220	51.3	4956206	0.00199	31.6
20–24	9111120	71.3	6496229	0.0024	28.1
25–29	9335700	75.82	7078328	0.00282	24.4
30–34	10205140	74.67	7620178	0.00378	20.7
35–39	11447350	76.55	8762946	0.00573	17.0
40–44	11454940	78.62	9005874	0.00859	13.2
45–49	10275640	78.03	8018082	0.01263	9.4
50–54	8995740	71.85	6463439	0.0191	5.9
55–59	7022870	59.79	4198974	0.03013	3.0
60–64	5668660	38.16	2163161	0.04802	1.2
65–69	5059830	17.17	868773	0.07376	0.4
70–74	4883870	8.77	428315	0.11263	0.0
75–79	4289810	3.1	132984	0.17444	0.0
80–84	3064520	0	0	0.27776	0.0
85–89	1836230	0	0	0.42268	0.0
90–94	876230	0	0	0.57447	0.0
95–99	302880	0	0	0.71	0.0
100+	65300	0	0	1	0.0

Sources: World Health Organization Life Tables 2000; US Census Bureau IDB demographic data 1990.

Table A.4.20 Reproducible capital

	US Reproducible Capital (billions of US dollars)	China Reproducible Capital (billions of US dollars)
1995 Stock	13,850.63	4,196.03
2000 Stock	17,655.75	6,356.76

Table A.4.21   Current account figures for US and China, 1995–2000

	US Current Account (billions of US dollars)					China Current Account (billions of US dollars)				
Year	Total	Goods	Services	Income	Transfers	Year	Total	Goods	Services	Income	Transfers
1995	−113.7	−174.2	77.8	20.9	−38.2	1995	1.6	18.1	−6.1	−11.8	1.4
1996	−124.9	−191.0	86.9	22.3	−43.1	1996	7.2	19.5	−2.0	−12.4	2.1
1997	−140.9	−198.1	89.8	12.6	−45.2	1997	37.0	46.2	−3.4	−11.0	5.1
1998	−214.1	−246.7	81.7	4.3	−53.3	1998	31.5	46.6	−2.8	−16.6	4.3
1999	−300.1	−346.0	82.6	13.9	−50.6	1999	21.1	36.0	−5.3	−14.5	4.9
2000	−416.0	−452.4	74.1	21.1	−58.8	2000	20.5	34.5	−5.6	−14.7	6.3

Sources: BEA (http://www.bea.gov/bea/di/home/bop.htm); National Bureau of Statistics of China (http://www.stats.gov.cn/english/); IMF 2002 *Balance of Payments Statistics Yearbook* Part 1.

Table A.4.22 Net holdings of international assets, 1995–2000

	US			China	
	Net Holdings of International Assets billion US Dollars (cost valuation)			Net Holdings of International Assets billion US Dollars	
Year	US cost valuation	US market valuation	Net External Position	Year	Net External Position
1995	−458.46	−305.84	−407.12	1995	−102.58
1996	−495.06	−360.02	−456.73	1996	−122.88
1997	−820.68	−822.73	−898.66	1997	−106.77
1998	−895.36	−1,070.77	−1,146.06	1998	−88.08
1999	−766.24	−1,037.44	−1,113.36	1999	−83.44
2000	−1,381.20	−1,581.01	−1,652.81	2000	−45.75

Sources: BEA (http://www.bea.gov/bea/di/home/iip.htm); Lane and Milesi-Ferretti (2007) (http://www.tcd.ie/iiis/pages/people/planedata.php).

Table A.4.23 Reproducible capital adjusted for international holdings

	US Reproducible Capital (billions of US dollars)	China Reproducible Capital (billions of US dollars)
1995 Stock	13,443.51	4,093.45
2000 Stock	16,002.94	6,311.01

Notes

1. China's official inflation estimates are lower than estimates from other sources (see, e.g., Young 2003), lending to uncertainty as to real GDP growth rates.
2. Human well-being depends critically on levels of health. Recent works by Nordhaus (2002) and Cutler and Richardson (1997) suggest that changes in health can have a value comparable to changes in GDP or other traditional income measures. In the near future we plan to integrate health in the comprehensive wealth framework described in this chapter.
3. We follow the general precedent of empirical studies in growth economics in measuring human capital by some function of the embodied years of education (see, e.g., Klenow and Rodríguez-Clare 1997, and Mankiw 1992) Of course, studies in human capital have also considered human capital as being formed by experience (e.g., Becker, Philipson, and Soares 2005), but the data we draw on have not made use of this or other refinements.
4. We abstract from uncertainty. For the purposes of determining sustainability over a short period of time, this is a legitimate approximation. However, for many policy purposes, uncertainty about the future should not be ignored.
5. In a similar spirit, Hamilton and Clemens (1999) introduced the term "genuine savings," where the modifier "genuine" distinguishes more comprehensive savings (savings that contributes to increased natural resource stocks as well as reproducible capital) from narrower, standard notions of savings.
6. In particular, Arrow et al. (2004) failed to take account of the capital gains to countries with large oil reserves. As a result, that study might have understated the sustainability of Middle East countries (see Table 2, p. 163, and discussion on p. 165).
7. Another way of looking at this is to consider the stock of knowledge as one form of capital. Then the growth in knowledge will be one form of investment, so that (8) does not have to be altered.
8. This adjustment of the investment flows for externalities is an approximation. A more refined approach (not taken in this chapter) would adjust as well the shadow values of each type of capital to account for the discounted value of the environmental damages (to the country owning the capital) caused by the use of that capital. Thus the shadow price of reproducible capital (including, in principle, durable consumer goods such as automobiles) would be reduced by the economic value of the health and disamenity costs imposed by particulate matter, sulfur dioxide, and other forms of pollution emitted, as well as, of course, the effects on global warming. If, over a given time-interval the amount of pollution increases, leading to greater environmental damages, the values of capital would be reduced to account for this change.
9. This data set underlies the estimates reported in Klenow and Rodríguez-Clare (2005).
10. The year 1995 was dropped such that we consistently use a 5-year period in all calculations.

11. See Ehrlich and Goulder (2006) for a discussion of potential limitations in existing comprehensive wealth studies, including biases from omissions of certain environmental damages.

References

Arrow, K. J., P. Dasgupta, L. Goulder, G. Daily, P. Ehrlich, G. Heal, S. Levin, K.-G. Mäler, S. Schneider, D. Starrett, and B. Walker, 2004. 'Are We Consuming Too Much?' *Journal of Economic Perspectives* 18: 147–172.

Becker, Gary, S., Tomas J. Philipson, and Rodrigo R. Soares, 2005. 'The Quantity and Quality of Life and the Evolution of World Inequality.' *American Economic Review* 95(1): 227–291.

Bolt, K., M. Matete, and M. Clemens, 2002. *Manual for Calculating Adjusted Net Savings.* Washington, DC: World Bank.

BP Statistical Review of World Energy 2005. Available on the web at http://www.bp.com/statisticalreview.

Cutler, David, and Elizabeth Richardson, 1997. 'Measuring the Health of the U.S. Population.' *Brookings Papers on Economic Activity: Microeconomics,* 217–271.

Dasgupta, P., and G. Heal, 1979. *Economic Theory and Exhaustible Resources.* Cambridge, U.K.: Cambridge University Press.

Ehrlich, Paul, and Lawrence Goulder, 2006. 'Is Current Consumption Compatible with Sustainability: A General Framework for Analysis and Some Indications for the U.S.' *Working paper*, Stanford University.

Fankhauser, S., 1994. 'The Economic Costs of Global Warming Damage: A Survey.' *Global Environmental Change* 4(4): 301–309.

FAO, 2006. *Global Forest Resources Assessment 2005.* FOA Forestry Paper 147. Rome: Food and Agriculture Organisation of the United Nations.

Hamilton, K., and M. Clemens, 1999. 'Genuine Savings Rates in Developing Countries.' *World Bank Economic Review* 13: 333–356.

Klenow, P. J., and A. Rodríguez-Clare, 1997. 'The Neoclassical Revival in Growth Economics: Has It Gone Too Far?' In B. Bernanke and J. Rotemberg (eds) *NBER Macroeconomics Annual 1997.* Cambridge, MA, and London: MIT Press, pp. 73–104.

Klenow, P. J., and A. Rodríguez-Clare, 2005. 'Externalities and Growth.' In P. Aghion and S. Durlauf (eds) *Handbook of Economic Growth.* Amsterdam: North Holland.

Lane, Philip, and Gian Maria Milesi-Ferretti, 2007. 'The External Wealth of Nations Mark II: Revised and Extended Estimates of Foreign Assets and Liabilities, 1970–2004.' *Journal of International Economics* 73: 223–250.

Mankiw, N. G., 1992. 'A Contribution to the Empirics of Growth.' *The Quarterly Journal of Economics* 107(2): 407.

Nordhaus, W. D., 2002. 'The Health of Nations: The Contribution of Improved Health to Living Standards.' *NBER Working Paper* No. 8818. March.

Nordhaus, W. D., and J. Boyer, 2000. 'Warming the World: Economic Models of Global Warming.' Cambridge, MA: MIT Press.

Nordhaus, W. D., 2002. The Health of Nations: The Contribution of Improved Health to Living Standards, in Kevin Murphy and Robert Topel, eds., *The Economic Value of Medical Research*, University of Chicago Press, Chicago.

Pezzey, John, 1992. 'Sustainable Development Concepts: An Economic Analysis.' *World Bank Environment Paper* No. 2, Washington, D.C.: The World Bank.

Stern, Nicholas, 2006. *The Stern Review on the Economics of Climate Change.* Office of Climate Change, HM Treasury, UK. Available at http://webarchive.nationalarchives. gov.uk/+/http://www.hm-treasury.gov.uk/independent_reviews/stern_review_economics_climate_ change/stern_review_report.cfm.

Tol, R. S. J., 2005. 'The Marginal Damage Costs of Carbon Dioxide Emissions: An Assessment of the Uncertainties.' *Energy Policy* 33: 2064–2074.

World Bank, 2006. *Where Is the Wealth of Nations? Measuring Capital for the 21st Century.* Washington, DC: The World Bank.

World Resources Institute, 1998. *World Resources 1998–1999: A Guide to the Global Environment.*

Young, Alwyn, 2003. 'Gold into Base Metals: Productivity Growth in the People's Republic of China during the Reform Period.' *Journal of Political Economy* 111(6): 1220–1261.

5
Counting Nonmarket, Ecological Public Goods: The Elements of a Welfare-Significant Ecological Quantity Index

*James Boyd**

1 Introduction

This chapter addresses a difficult, important, and long-standing problem in national income accounting: How do we capture the welfare contributions of nonmarket, public environmental goods?[1] As I argue here, the key to this endeavor is the construction of a welfare-significant ecological quantity index (WSEQI). Quantity indexes and their price index counterparts are the core of any national income or product account (NIPA).

What quantity units are we to count when markets are not present, meaning that there is no easy way to track or define what is being used, consumed, or enjoyed? Note that market goods usually come in conveniently predefined quantity units—the cars, washing machines, haircuts, and restaurant meals consumers buy every day. In conventional income accounts these goods and services are the building blocks of a quantity index. We can simply ask: How many cars and haircuts were consumed during this period? But for natural capital, its many public, nonmarket goods and services are not defined, provided, and priced by markets.[2] This means, of course, that we lack the market artifacts—units sold, prices paid—so useful to accountants of the market economy.

Not only do we have a missing prices problem—How do we infer value when market prices are absent?—we also have a missing quantities problem. Stated more carefully, we need to replicate the market's ability to define units of consumption, but do so in a nonmarket setting. Accordingly, economists should be involved in defining ecological quantity accounts and work with natural scientists to depict nature in a way useful to utilitarian analysis.[3]

Here, I advocate using principles from economic accounting, welfare economics, and environmental valuation to define the nonmarket units that should be used in a quantity index of ecological goods and services. My goal is to mimic the household-level foundations of conventional NIPAs, but extend them to the nonmarket natural economy. In this chapter, I describe the underlying theory and show how it can be used to define a practical, empirical strategy for constructing a WSEQI.[4]

Capturing nature in a comprehensive way is a tall order. Ecological systems are complex, with an uncountable number of components interacting nonlinearly. Is it realistic to think we can capture such a complex system in a practical account? It depends on the goal, of course. If the goal is comprehensive knowledge of nature, that is impossible. But accounting systems serve narrower ends. They provide a rough, but valuable, guide to the more complex systems they describe. The conventional economy is also complex, multidimensional, and nonlinear. We do not look to gross domestic product (GDP) and other NIPAs for the complete truth about our economies. Instead, we look to them as important signals of our welfare. Nature's nonmarket contributions to our well-being deserve a similar set of signals.

1.1   The goal of the index

The index described here is intended to create a "welfare-significant" measure of nonmarket ecological consumption. Such an index has several characteristics: First, it is anthropocentric and rooted in utilitarian economics. Second, the index is an account, not an indicator system.[5] Accounting systems rely on "identities" to facilitate and discipline measurement. At the firm level, double-entry bookkeeping is an example. At the national level, so is the definition of GDP.[6] Accounting identities facilitate aggregation and comparison of the components of an index in ways that indicator systems do not. Despite being constructed from myriad components, GDP can be reported as a single value because accounting identities discipline its construction. The political and social influence of NIPAs derives largely from the fact that they are rule-based accounting systems.

What accounting identities are binding in a WSEQI? They are the same as those in a GDP-like NIPA, with one difference. Because the focus is on nonmarket goods and services, "virtual" prices and income are important. Although the measurement of virtual prices and income is not my focus here, the weights eventually attached to the quantities I define in this chapter should be the virtual prices that emerge in a general equilibrium constrained by the sum of real and virtual income.[7] An additional set of constraints arises relating to the distinction between intermediate and final goods. I address this in detail in Section 3.

1.2 Relationship to other environmental accounting approaches

A full review of the many alternative approaches to environmental accounting is beyond the scope of this chapter.[8] The ecological quantity index I describe in the sections that follow is related to, but different from, both material accounts and NIPA-like environmental accounting systems, often loosely referred to as "green GDP" or "green national accounts." As I will argue, a WSEQI comprises physical material measures. A WSEQI, however, is not a "material account" as that term is used in national accounting. I describe the difference between a WSEQI and a material account in detail in Section 3.2.

Most green income accounts start with the presumption central to this chapter, that nature's contributions to welfare should be measured. In practice, however, existing green accounts differ from the WSEQI described here. For very good reasons, existing accounts focus on "near-market" clues to nature's value.[9] Examples include economic damages arising from air pollution (using health costs), the valuation of timber stocks (using land values) or fisheries (using the value of commercial harvests), and the value of water (for hydropower generation). China's environmental accounting system, for example, focuses on near-market accounts—such as its accounts of environmental expenditures—along with aggregate assets such as fish populations, forests, water, and minerals (The World Bank 2006). It is natural for accounting systems to focus on near-market assets, goods, and services because the near market is where credible price estimates are most practically derived.

The downside of near-market accounts, of course, is that the scope of these accounting systems is relatively narrow. They ignore the environment's public, nonmarket goods and services, focusing instead on goods and services that can be credibly priced. The WSEQI makes a different trade-off. It seeks comprehensive measurement of public, nonmarket goods and services, but in doing so it moves away from the market and the prospect of easily derived virtual prices. We can see, then, that green national accounts and a WSEQI share the same motivation but are applied to different streams of goods and services. They are complements, not competitors.[10]

Although ecological income accounts are recognized as important, these are in an early stage of development.[11] No practical example exists.[12] Ecological accounts are beginning to be developed, but these are not income accounts in an economic sense.[13] As a result, for China—or for any other country—the accounts described here are not necessarily the most practical first step in efforts to make NIPAs reflect environmental costs and benefits. But nature's nonmarket contributions to welfare may be at least as large as its near-market contributions (The World Bank 2006). It is important and interesting to understand the sheer magnitude of that virtual natural economy.[14]

2   Quantities versus weights

Economic measurement of nonmarket benefits typically ignores the distinction between quantity q and value p. What matters is the product of the two, $p \cdot q$, the social benefit of an air-quality improvement, for example. Welfare-significant accounting systems require a precise, consistently maintained delineation of the quantity measured and the price or other weight attached to the quantity.

2.1   The index number problem

Accounting measures, by their very nature, distinguish between quantity and price. We can think of economic accounting theory as a search for ways to factor the benefits of production into (1) their two core components, price and quantity, in a way that is (2) logically and economically consistent. This challenge is what Irving Fisher called the "index number problem."

With quantities and prices clearly differentiated, one of the quantity/price sets can be held constant. Real GDP, for example, is a measure of quantity, in which prices are held constant over time. With prices held constant, movements in the output index meaningfully describe changes in quantity produced and consumed. If prices are not held constant, the interpretation of the index is muddied: Is an increase in the index evidence of changing prices or changing output? Accordingly, economic accounting systems require a clear and consistently maintained distinction between q and p.

It is interesting that the distinction between q and p gets so little attention in environmental economics. Nonmarket goods and services do not come in convenient packages, where q is defined by the market. So how do nonmarket environmental economists think about this? The truth is, they don't, largely because they haven't needed to.

Environmental economics is often called on to analyze the following type of issue: What are the benefits of a tighter air quality standard in Los Angeles? To answer this question, all that matters is the comparison between $p \cdot q$ before the policy is implemented and $p \cdot q$ after it is in place. It is unimportant what part of the benefit is considered an improvement in output, versus an increase in the value of the output. For example, is the number of people in Los Angeles, n, considered part of the value of the benefit or part of the quantity? It could be either.[15] And for many questions in environmental economics, it doesn't matter.

But it matters a lot to economic accounting. Because output and price measures are constructed separately, their units must be distinct and consistently applied.

2.2   What is quantity and what is value?

If q and p are to be distinguished, how should this be accomplished? What principles, if any, should be applied? Because nonmarket ecosystem goods and services do not emerge from factories and are not sold in markets, defining and measuring their "units of account" requires theoretical and empirical innovation. The conceptual distinction between ecosystem services and their value is often surprisingly difficult to make.

Accountants use the term "goods and services" to denote the quantities measured in accounts. This same convention is applied here: Ecosystem goods and services are the quantities to be measured. Prices, virtual and otherwise, are the weights applied to the goods and services. A problem arises, though, in that the term "ecosystem services" has no consistent definition in environmental economics.

Next, I offer a brief mathematical synopsis of Banzhaf and Boyd (2005), using a simple production technology to convey the semantic and practical distinction between q and p. The model features a final good F, and two inputs, capital K and an ecosystem good E. The final good and the inputs have prices P_F, P_K, and P_E, where P_E is a virtual, not observed, price. The production function is some $F = F(K, E)$.

Two common approaches are taken to the depiction of the ecological input. First, the value of the ecological input can be derived from the value of its productivity with respect to the final good times the price of the final good.

$$P_E = (\partial F / \partial E)P_F. \tag{1}$$

Second, using the production function itself and some rearrangement of terms, the ecological input's value can be expressed via the other input's role in production. Specifically,

$$P_E = \frac{\partial F/\partial E}{\partial F/\partial K} \cdot P_K, \tag{2}$$

where the ecological input's value is derived from its substitution relationship with the capital input times the price of the capital input.[16]

Now consider a marginal change in the ecological input, ∂E. The total value of that change can be expressed in three different but equivalent ways.

$$P_E \cdot dE = (\partial F/\partial E) \cdot P_F \cdot dE = \frac{\partial F/\partial E}{\partial F/\partial K} \cdot P_K \cdot dE. \tag{3}$$

Do these expressions give us a clear guide to the "quantity" (the ecosystem good or service) and the "value" of the good or service p? Both in principle and practice, the answer is no.

For example, individual chapters in a well-known environmental economics text (Kopp and Smith 1993, Chapters 2, 7, and 14) define ecosystem services in three completely different ways. One defines the service as the total value—$q = P_E \cdot E$ in our example. Another refers to the service as the contribution of the environmental input to production of the final good as E changes. Denote this as $q = \Delta F(K, E)$.[17] Yet another refers to the service as the environmental input itself, $q = E$. In a total benefit framework, each of these definitions is consistent with the identity in equation (3). In other words, none of the definitions is "wrong." From an accounting perspective, though, the lack of clarity and consistency is problematic.

So which definition of q—the ecosystem services to be counted—is preferable? We can immediately rule out the first because it defines q as the product $P_E \cdot E$, instead of something that can be decomposed into distinct quantity and value components. The distinction between $q = \Delta F(K, E)$ and $q = E$ is more subtle. Both definitions permit multiplication by a price to achieve a total value.[18] In that sense, both work from an accounting perspective. But the latter definition, where the ecological input is the quantity measure ($q = E$), is preferable.

This definition eliminates the need to understand economic production functions for the purposes of developing the quantity index. This means, though, that the economic production functions depicted in equation (3) must be captured on the value side of the accounting system. If clean water is a quantity to be measured in the index, estimation of the value of that water is where substitution relationships and knowledge of the way they translate into final production are captured.[19] Absent direct knowledge of P_E, we cannot avoid the complexities arising from joint ecological and nonecological production. But there are practical reasons to move those complexities to the valuation side of the ledger.

If we use $q = \Delta F(K, E)$, that definition of quantity can obscure, rather than clarify, underlying ecological changes. If all we observe and count is an increase in final production, we will not know whether that increase results from changes in the ecological or the nonecological inputs. In fact, innovation or other changes in production may lead to higher production levels even if ecological inputs are declining in availability or quality. This is an undesirable property for an ecosystem services quantity index. An ecological quantity index should tell us about ecological conditions; it should not require economic interpretation. Consequently, $q = E$ is preferable.

There are other reasons to prefer $q = E$. The measure of value associated with $q = \Delta F(K, E)$ is P_F. If that price is available, we would have a practical reason to prefer its associated quantity measure. But there will be no such price for nonmarket goods and services. Also, ecological measures $q = E$ are concrete and intuitive, they make sense to noneconomists, and they are in the empirical realm of the natural sciences. Economists should have something to say about which elements E are measured. But economists themselves will not do any of that measuring. In addition, ecological measures $q = E$ already have a close analogue in national accounting systems. They resemble the material accounts already observed in many international systems.

In summary, an index composed of concrete, physical, ecological quantities and qualities is the appropriate place to begin a system of ecological public good accounts. Moreover, such an index—because it is "material and physical" instead of "economic"—provides a natural point of collaboration between the natural and social sciences. I develop this point in more detail in Section 4.

Keep in mind that the quantity index advocated here, although material and physical in composition, is not equivalent to material accounts as they are understood by the international accounting profession (Smith 2007). I develop the distinction in more detail in the sections that follow.

3   The elements of a WSEQI

In this section, I describe physical ecological measures—the quantity index—but they are physical measures consistent with an economic, rather than materials-based, approach to accounting. I use the word "quantity" as shorthand for all of the following: countable biophysical features (e.g., land cover types and species populations); biophysical qualities (e.g., particulate or toxics concentrations); and stochastic depictions of quantities and qualities (e.g., hydrographs).

Previously, I advocated using ecological inputs themselves as the elements to be valued in a WSEQI. But as I pointed out earlier, nature presents us with an uncountable number of such inputs. So, to comprehensively depict its contributions to well-being, must we count all nature's features and qualities? The answer is no. Much as GDP does not count all the units exploited and traded in the market economy, an ecological quantity index need not count all the elements of nature. In both cases, a utilitarian approach to accounting allows us—requires us, actually—to focus on "final" units of consumption. The intermediate inputs to those final units are explicitly excluded from an economic quantity index. If they weren't, the quantity index would double count numerous elements of consumption, which violates the accounting identities so central to an economic index.

What do "final goods" mean in an ecological context? In the absence of markets, is final ecological good a concept that can even be interpreted? That is the main subject of this section, although I also address several other issues, including the relationship of the WSEQI to material and asset accounts.

3.1   Biophysical final goods

In Section 2, I argued that the ecological quantities best suited for an economic index are biophysical inputs (E) to economic production. I then introduced the notion that we should count final goods and services. This construction suggests that the things we should count are both inputs and final goods—an apparent contradiction. To clarify, let me restate the role of an ecosystem account in a larger system of economic accounts. The goal of the quantity index described in this chapter is to comprehensively describe both market and nonmarket ecological "exports" to the economic system. Semantic confusion can arise because two distinct systems are in play. First is the biophysical system (nature). Second is the economic system that translates biophysical inputs into economic benefits. The terminological difficulty crops up because the ecological quantity index counts the biophysical system's final goods and services. But these final biophysical quantities then appear as inputs to GDP or other economic accounts. The final goods and services described here, then, are final only in a biophysical sense. They need not be final in an economic sense.[20]

To define biophysical final goods, let's start with how these goods are defined in existing economic accounting systems. Two principles help define what it means to be a final good. First is the need to avoid double counting. Second is the importance of consumer choice to valuation.

First consider the issue of double counting in a welfare-significant account. If we count both cars and the steel used to make them and then weight cars and steel by their market prices, we will have double counted the value of the steel because the steel's value in car production is embodied in the value of the car. In calculating GDP, final and intermediate goods are distinguished in the following way. If a good or service's value adds to the value of a good or service subsequently sold in the market, it is an intermediate good. Otherwise it is a final good. Returning to my example, cars are final goods. The labor, leather, steel, and human capital required to make the car are intermediate goods.

The concept of double counting illustrates a fundamental difference between material and economic accounts. Material accounts can resemble input–output models, where the goal is to track the life cycle of physical resources so that resource demand, waste, and externalities can be clearly identified. In general, double counting is not an issue in material accounts because a goal of such accounts is to track the transformation of a resource as it moves through the economic system.[21] For example, in material accounts, wood will appear in many different forms throughout the accounts—perhaps as standing forest, raw timber, pulp, waste, and finished lumber. In contrast, where economic accounts are concerned, double counting violates the underlying economic identities that constrain the account.[22]

> Thus, it deserves emphasis that a WSEQI is material in nature. But material and physical accounts—as commonly understood—are not themselves a welfare-significant quantity index.

Second, the importance of consumer choice is taken for granted in GDP, but should not be taken for granted in environmental and other nonmarket accounts. When counting market goods, those goods are subject to consumer choice by definition. In fact, the value of final goods is revealed by consumer choices in the market. In a nonmarket context, the point at which consumers make choices can be less clear. GDP tends to count items that are concrete and subject to tangible consumer (market) choices.[23] If we are ever to attach welfare-significant weights to ecological inputs, the quantities we count should have the same properties.

Given that the ultimate aspiration of the index is to weight ecosystem goods and services with their virtual, nonmarket prices, the quantity index needs to be composed of units subject to valuation. The only way to ever estimate these virtual prices is to come as close as possible to the point at which people reveal those virtual prices through choice. This is the rationale for viewing ecological accounting units as the final inputs to "home production." Another way of putting this is

> The final biophysical units used in an ecosystem quantity index should be the ecological features, quantities, and qualities that are directly combined with other (nonecological) inputs to produce market and nonmarket benefits.

With units such as these, virtual prices can then—in principle—be derived via analysis of the ecological inputs' contribution to market outputs, as in equation (1), or by knowledge of substitution possibilities, as in equation (2).

3.2   Illustrations

In practice, the procedure is to first identify utilitarian benefits that require ecological inputs, then identify the ecological final goods used as inputs to those benefits. Consider the following incomplete list of benefits, associated final ecological inputs, and nonecological substitutes.

Recreational angling. As a recreational experience, angling benefits arise from a combination of ecological final goods such as lakes, streams, riparian land cover, and fish populations present in the water bodies. Benefits also arise from nonecological inputs such as capital goods (rods, lures, boats, and docks) and human capital (expertise).

Note that the "number of fish caught" is not the quantity measure we seek in this index. Why? Because the number of fish caught is a function of ecological and nonecological inputs. The better the equipment and the greater the skill of the angler, the more fish are caught—independent of the state of the underlying ecosystem. This example reflects the principle that the index's quantities are those relevant at the point of possible substitution to nonecological inputs.

Also note that there are multiple ecological inputs to the ecological final goods identified here. Fish populations relevant to angling depend on the populations in those species' food chains, for example. But food chain species should not be counted as final goods in this context (the benefits of recreational angling).[24] Instead, their value is embodied in the bass, trout, salmon, or other population targeted by anglers.

Flood damage mitigation. Reduced frequency and severity of floods is a utilitarian benefit to which ecological inputs can contribute. Wetlands, in particular, absorb and slow flood pulses. To a lesser extent, other natural land cover types do so as well. The quantity of wetlands, then, is a final ecological input to the provision of flood damage mitigation.

The virtual price of wetlands (the value side) could, in principle, be derived either from knowledge of the direct effect of wetlands on property and other damages, or indirectly via the effects of nonecological substitutes such as dikes, dams, or other forms of property protection.

Pollination of commercial crops. Native species play an important role in pollinating commercial crops and can directly influence crop yields. As a result, the presence and density of native pollinator populations is an important input to commercial agriculture.

Again, commercial harvests are not the appropriate ecological quantity measure because harvests are the product of both ecological and nonecological inputs. The virtual price of pollinator populations can, in principle, be inferred by controlling for the presence of other inputs (and thus learning about the pollination–harvest production function) or from the prices and substitutability of commercial pollination services.

Public health damage mitigation. Air, soil, and water quality are the appropriate quantity measures in this context. Because reductions in acute health events, morbidity, and mortality result from combinations of ecological stressors and nonecological inputs, such reductions are not the proper quantity measure. Substitutes include medical interventions, filtration, and damage avoidance actions.

Aesthetic benefits. Ecological features that directly give rise to aesthetic benefits tend to be related to land cover types. Undeveloped terrain, open water, and mountain areas are relevant final quantities here. So too are certain types of air and water quality. Clear water and clear air, apart from their other benefits, can also contribute to aesthetic benefits.

There may be no clear substitutes for certain aesthetic features of the natural landscape. Economic production of these benefits, however, often requires complementary investments in time and access. After all, a beautiful view has no aesthetic value if it cannot be seen.

Stewardship benefits. Existence, nonuse, bequest, or stewardship benefits arise from altruistic and ethical motivations. But there are clear quantity measures of these benefits, namely the species, wilderness, and natural features to which we attach existence value. For example, a raw count of viable global species—perhaps weighted by the charisma of individual species—is a quantity measure of species existence benefits.

Uniquely, there are no nonecological substitutes for stewardship benefits. Moreover, no economic production function intervenes between the ecological final goods and the benefits arising from them.

I should emphasize that the examples I give here are not exhaustive of the benefits to which nature contributes.

3.3   Final goods are benefit-contingent

Although the principles and constraints imposed by accounting identities are central to an economic accounting system's power and validity, they can lead to confusion. A prime example is the way in which final goods are benefit-contingent. This means that a good can be final in the provision of one benefit and not final in the provision of a different benefit. Consequently, many of the final goods identified in my illustrations are final goods only in that particular context. Ecological inputs, then, will switch back and forth between final and intermediate, depending on the benefit being accounted for.

Consider a hillside forest and two different kinds of benefit: aesthetic benefit and public health damage mitigation. Aesthetically, people with visual access to the hillside directly enjoy the forest's physical features, so those features should be counted as a beauty-related final quantity. In terms of public health, forests may sequester pollutants. In this benefit context, however, the forest is an intermediate, not final good. The final, public-health-related good is the air quality itself. The forest has a positive, but intermediate, impact on that final good. Accordingly, from an accounting perspective, the forest is both final and intermediate.

Other examples of this phenomenon abound. Wetlands should be counted as final goods for flood protection but are intermediate goods when they lead to improvements in drinking water quality (in which case the drinking water quality is the final good). In a conventional GDP context, the same thing happens. Tomatoes, onions, lettuce, and ground beef are counted if purchased in a grocery store, but are not counted if sold as a McDonald's hamburger.

3.4 Counting discrete ecological goods and services, not ecological assets

An ecological index cannot be welfare-significant if it cannot disaggregate ecological inputs along spatial and temporal dimensions. For example, what is the value of U.S. water resources in 2007? This question cannot be meaningfully answered without more detailed knowledge of the location and timing of the waters' availability.

To date, most green accounting systems have adopted a different approach, which we can call an "aggregate" approach to environmental inventories. Mineral and forest resource accounting, for example, often uses national aggregates, such as the total supply of harvestable timber, copper, and water resources.[25] In these applications, an aggregate approach to accounting makes sense. Aggregate measures of these kinds of commodities are entirely appropriate because lumber and copper are fungible (homogeneous, transportable, and storable). Second, it makes sense to make use of aggregate asset prices when they are available, as they are in the case of many commodity-type natural resources. Most existing green accounts exploit the availability of commodity asset prices in just this way, making aggregate asset measures the natural, corresponding quantity measure.

An aggregate approach to ecological nonmarket value, though, is unsatisfactory.[26] Ecological inputs are not fungible in the way that most economic goods and services are. Rivers and forests cannot be shipped across state lines. Similarly, most ecological inputs cannot be accelerated or inventoried across time (note, though, that this is precisely the purpose of reservoir management). If welfare significance is the empirical goal, the quantity units in the index must reflect a fine-grained sense of space and time.[27]

This is particularly true given the ultimate goal of valuation. Like any benefits, environmental benefits are a function of scarcity, substitutes, and complements. Environmental benefits are often not fungible precisely because substitutes and complements in the economic production function are themselves not fungible.[28] If a beautiful vista is to yield social value, people must have access to it. In other words, the vista must be spatially "bundled" with infrastructure—roads, trails, and parks—that are themselves not transportable.

Recreational fishing and kayaking require docks or other forms of access. Substitutes for a given recreational experience depend on a recreator's ability to reach them in a similar amount of time. For this reason, the location of nonfungible substitutes is important. The value of surface water irrigation is a function of the location and timing of alternative subsurface water sources. If wetlands are plentiful in an area, a given wetland may be less valuable as a source of flood pulse attenuation than it might be in a region where it is the only such resource. Accordingly, ecological goods are not fungible and neither are the substitutes and complements necessary to their eventual valuation.[29]

Many market goods are bought and sold as assets (e.g., real estate, financial products, and firms), but nonmarket public goods are unlikely to ever be sold this way. Even when they are, the observed prices should be treated with suspicion, given the necessary role of government in assigning quasi-property rights and setting prices.[30] Ecological accounting, then, should not expect aggregate, asset-type market prices to be empirically relevant. Instead, the opposite is true. Public good, nonmarket ecological assets are much more likely to be valued by "building up" the value via valuation of the services flows arising from the asset.[31]

This mind-set is very different from that expressed in the System of Integrated Environment and Economic Accounting's (SEEA) discussion of ecological accounting (although note that the SEEA offers no concrete proposals for the way in which ecological accounting should occur). According to the SEEA, "it is not generally the components of ecosystems that benefit humans, but the systems as a whole."[32] This is both philosophically debatable and practically unhelpful. Society surely benefits from the ecological system, but the same can be said of the market system. Nevertheless, when we account for the market economy, we do not value the system as a whole. Instead, we construct the system's value from its discrete components.

3.5 Ecological quality

How are goods and services of different quality handled in economic accounting systems? And how should they be handled in an ecosystem index? Clearly, because quality matters to welfare, it should be captured by the accounting system. Interestingly, quality differences can appear on either the quantity side q of the accounts or on the value side p. Which way it is done is a matter of practical choice. The choice is not based on a black-and-white economic principle.[33]

Consider two different kinds of wetlands, one that significantly absorbs flood pulses and one that doesn't.[34] Ideally, we should treat these as distinct goods, account for them separately, and assign them distinct virtual prices. Clearly, though, there are practical limits to the level of disaggregation that can take place in a working set of accounts. The alternative is to count at a higher level of aggregation and have the corresponding virtual price reflect the underlying quality heterogeneity. The existing NIPAs routinely confront this kind of choice.

Imperfect quality differentiation and adjustment is a recognized weakness of existing NIPAs. Consider the way in which U.S. accounts treat changing product qualities over time. GDP counts computers, but because of technological innovation a computer in 1990 is clearly not the same as one in 2005. Unfortunately, adjustment for quality differences resulting from innovation or even certain basic product characteristics (shouldn't Apple computers be a different product category than PCs?) creates measurement difficulties. Currently, the national accounts selectively apply hedonic quality adjustments only in certain product categories.

The practical measurement of ecosystem goods and services will raise similar issues. But an ecological quantity index—like an economic quantity index—should differentiate goods and services by their quality to the greatest extent practicable.

4   Ecological prediction, sustainability, and depletion adjustments

It is possible—and many fear—that our current human footprint is robbing our children of future well-being. The fear is expressed economically as a concern that we are overconsuming, eating into our natural capital's principal, and consuming natural resources faster than they can reproduce or regenerate. A central tenet of welfare-significant economic accounting is that unsustainable consumption should create a debit in the current account. A depletion-adjusted account is a fact-based, rigorous way to measure sustainability or its lack. An ecological index that does not grapple with depletion will be unsatisfying to environmentalists, ecologists, and economists alike.

The problem, of course, is that depletion analysis is difficult, since it requires us to know the complex, underlying causes of depletion (El Sarafy 1989). Even in the case of subsoil assets (e.g., coal, natural gas, and copper), the depletion relationships used in accounts are relatively crude and subject to ongoing debate. Ecological depletion will be an even taller order in this respect. How do air emissions from coal-powered plants affect water quality throughout the airshed? How does a residential subdivision affect species populations in surrounding counties? Today, we know that these questions are important, but we are far from the empirical consensus needed to make depletion adjustments in practice.[35]

A richer understanding of ecological causality is necessary if we are to create economic accounts that capture depletion phenomena. Causality will empower prediction, and prediction will empower the incorporation of depletion costs into current economic accounts. Then, the practical question becomes: How can we promote a richer understanding of ecological causality? One answer is the construction of an ecological quantity index such as the one I describe in this chapter.

A quantity index is a snapshot of consumption. In the GDP context, the quantity index counts the level of goods and services consumed in a given period. Similarly, an ecological quantity index would count natural features and qualities occurring in a given period. Directly, this doesn't tell us anything about the direction or rate of change of those features and qualities. Enough snapshots over enough time, though, create a multilayered, national-scale, biophysical, time-series database, which would be a huge boon to the biophysical sciences. It would allow us to analyze ecological causality at the landscape level in a way that is currently impossible.[36] Depletion adjustments to a current account can occur only if there is an empirically defensible record of ecological causality to justify them. The early years of data collection won't tell us much about the future. But with each passing year prediction will improve, as will our ability to adjust the accounts intertemporally.

5   The political economy of environmental information

If voters in coming elections are asked, "Are you better off environmentally today than you were four years ago?", their answers will be anecdotal and impressionistic, not based on a comprehensive set of hard facts. This is true in part because no set of national statistics summarizes our environmental well-being. When the market economy suffers from inflation, unemployment, or negative growth, our society can largely agree on these facts because they are a culturally, governmentally, and scientifically sanctioned set of measures: the NIPAs. How do we make available to voters a similarly credible set of statistics that reveal the state of our environmental well-being?

This chapter makes a small contribution to the theory and practice of environmental accounting by focusing on one of its most difficult challenges: the measurement of ecological, nonmarket public goods in a way consistent with economic accounting principles. Here, using ideas from both environmental valuation and national income accounting, I advocate constructing a WSEQI. I also identify a set of practical principles to guide the identification of quantity units suitable for such an index.

Eventually, such an index will need its corresponding value index (the virtual prices used to weight quantities). But as I show in this chapter, focusing on the valuation side of the ecological index problem puts the cart before the horse. We can value only what we can consistently count. Those quantity units should be constructed so that they are amenable to valuation. For this reason, I stress issues like temporal and locational specificity, avoidance of double counting, and quality adjustments. But until we debate and develop "a theory of quantities," and marry the theory to concrete methods of measurement, the valuation problem will remain academic, not practical.

Economists have naturally focused on the valuation side because that's where their skills lie. But economic principles are also important to the quantification side. More economic thinking should be devoted to the material and physical accounts economists normally ignore. Why? Because physical accounts are a precondition of welfare-significant ecological accounting.

To conclude, it is worth reflecting on the practical, institutional issues raised by the need for environmental information. What are the prospects for an "environmental information movement" that could advance the practice of ecological accounting? National accounts take decades to develop, even when created in response to an institutional mandate, and they require financial and political support. The methodologies of existing accounts, such as U.S. GDP, are still actively debated and corrected even 75 years after they were first constructed (Carson 1975). But even so, environmental accounts have been slow to develop, although they have been advocated for decades (see, for example, Ayres and Kneese 1969; U.S. BEA 1982; United Nations 1984; El Sarafy 1989; Peskin 1989; and Repetto et al. 1989).

The current lack of comprehensive environmental accounts can be explained, of course, by country-specific factors. But several generic barriers are worth noting—all of which can be overcome in the years ahead. The first barrier is the need for coordination between the natural and social sciences. Conventional NIPAs do not rely on the natural sciences to any great degree. Environmental accounts demand biophysical analysis and measurement. Alone, ecologists can study and report on the characteristics of nature. Alone, economists can opine on the economic value of nature to households and the market economy. But national environmental accounting requires complementary, coordinated activity from the two realms. The accounting approach I describe in this chapter aspires to precisely this kind of integration.

Fortunately, the two disciplines are now working more closely than ever before. Ecologists increasingly see nature's broad contributions to economic well-being as a subject for ecological study (Daily 1997; Kremen 2005; Millennium Ecosystem Assessment 2005). Conservationists and environmental trustees also increasingly view economic arguments as useful to their mission. Likewise, economists have become much more sensitive to and skilled at analyzing nature's goods and services, including those that resist traditional economic analysis (Heal 2000).

The second barrier is the cacophony that results from an overabundance of unrelated environmental indicators, performance measures, and statistics. It is hard to argue against any particular effort, but as a whole these competing measures undermine the power of all. Imagine what would happen if we had several competing sets of economic statistics. None would be trusted and all would be used opportunistically to serve the political ends they support.

Clearly, numbers that appear to be provided by biased sources lose their political and economic power.[37] The global experience with NIPAs clearly demonstrates that credible statistics require a combination of institutional independence and accountability. Independence is achieved by housing the effort in politically and bureaucratically insulated institutions. As a corollary, environmental accounting should probably not be housed in an environmental agency. An agency responsible for environmental management should not also be the keeper of the books. Accountability is achieved via centralization, which makes information providers more accountable to the political process, not less.[38] Moreover, national-scale data collection is expensive and likely to have scale economies. This is another virtue of centralization.

A third barrier to national environmental statistics is the government institutions charged with protecting and managing our environment. Although they create a lot of information, U.S. agencies such as the U.S. Environmental Protection Agency (EPA), the National Oceanic and Atmospheric Administration (NOAA), the U.S. Army Corps of Engineers, and the Department of Interior have not to date been sources of consistent, comparable, and national statistics. This is largely because the agencies serve distinct and limited missions and mandates, and each collects information related to its narrow sphere of authority. There is no bureaucratic incentive to harmonize statistics because no agency has the authority or budget to comprehensively track environmental outcomes. Federal environmental agencies may actually oppose the centralization and independence of environmental statistics. Finally, government environmental agencies are accustomed to a culture of political consensus, and rightly so. But NIPAs require more than political consensus. They also necessitate scientific consensus based on vigorous intellectual debate that is insulated from bureaucratic politics.

A final barrier to be confronted is the fact that environmental statistics benefit no one in particular, but everyone in general. In other words, environmental statistics are themselves a public good. As Olson (1965) noted 40 years ago, the public goods least likely to be provided are those where the costs are concentrated and the benefits are widely shared. The benefits of environmental statistics are certainly diffuse, not concentrated.

The science of ecological measurement is advancing rapidly. So too is the idea of concretely measuring nature's contributions to our well-being. A political mandate for comprehensive national environmental accounting, though, is lagging. When that mandate arrives, a WSEQI will hopefully be pursued alongside other accounting tools. If, as many believe, nature is so central to our well-being, we need better information on the state of our natural wealth. A WSEQI is one important way obtain this information.

Notes

*Stanford University and Resources for the Future (RFF).

1. For histories of the idea and its role in national accounting, see Mäler (1991), Nordhaus and Kokkelenberg (1999), and Heal and Kristrom (2005).
2. Even when nature's goods and services aren't public goods, the private consumption of nature often has ancillary consequences for the scale or quality of public goods. In addition, many natural resource markets for private goods are heavily distorted by inefficient regulatory regimes, rendering the market information that they provide suspect.
3. Banzhaf and Boyd (2005), Boyd (2006), and Boyd and Banzhaf (2007) develop this theme and analyze concrete ways in which economics and ecology can productively interact.
4. Practicality and a sound theoretical base are the prerequisites for a successful accounting system. As Heal and Kristrom note (2005, p. 1211), the way forward is "to find an even happier marriage between theorists and empiricists in green accounting."
5. All accounting systems are indicator systems, but the reverse is not true. The distinction is that accounting systems are constrained by their structure in a way that measurement systems are not. Ecological accounting to date takes the form of indicator systems. For an example, see Binning et al. (2001).
6. Double-entry bookkeeping means that each transaction results in at least one account being debited and at least one account being credited, with total

debits equal to total credits. GDP is defined as the sum of consumption, investment, government purchases, and net exports.

7. I should emphasize that GDP-like NIPAs and the accounting system described here are not, and can never be, a measure of welfare itself. Instead, the indicators they produce are best thought of as practical approximations of welfare.

8. For descriptions of existing environmental accounting initiatives, see Hecht (2000, 2005), Haas et al. (2002), and Heal and Kristrom (2005).

9. As an input to marketed goods, nature's value is partially captured by NIPAs, although nature's specific contribution to the value of the market goods and services is not extractable as an independent set of measures.

10. Arguably, neither of these should be called green GDP because that term suggests a comprehensive measure of market and nonmarket well-being.

11. According to Hecht (2000, p. iii), existing environmental NIPA efforts "include neither meaningful adjusted macroeconomic indicators nor the value of non-marketed environmental goods and services."

12. Describing the recent System of Integrated Environment and Economic Accounting (SEEA) handbook on environmental accounting Smith (2007, p. 597) observes that "ecosystem accounts are in their relative infancy and are presented more by way of suggested avenues for exploration in the handbook than as clearly worked out recommendations."

13. See Weber (2007) for discussion of current ecological accounting in Europe.

14. The U.S. National Research Council (1999, p. 23) states the mission thus: "We must not forsake what is relevant and important merely because it presents new problems and difficulties.... We must endeavor to find dimly lit information outside our old boundaries of search, particularly when the activities are of great value to the nation."

15. If the quantity is defined as a per capita air-quality improvement, the number of people benefiting would increase the weight (p) given to the quantity. If, on the other hand, we defined the quantity as the change in the total amount of human exposure, n would increase the quantity q, not the weight p.

16. Algebraically, this is derived from the producer's tangency condition $(\partial A/\partial E)/(\partial A/\partial K) = P_B/P_K$.

17. For example, if the ecological input goes from E^1 to E^2, $q = F(K, E^2) - F(K, E^1)$.

18. In the former case, the final good's price P_F pertains. In the latter case, we require either knowledge of the virtual price P_E or knowledge of the production function and P_K.

19. The production functions described here should be thought of as economic production functions, where ecological inputs are combined with noneconomical inputs to produce outputs that are consumed or enjoyed by society. A distinct set of "biological" production functions describes the way in which ecological outcomes arise from ecological and social conditions.

20. Some final ecosystem goods are also final economic goods. An endangered species, for example, is a final good in both a biophysical and economic sense because the existence benefit of the species requires no intervening, noneconomic inputs to yield an economic benefit.

21. "Physical accounts suffer from one major drawback—at least in the eyes of users who view the world through an economic lens: they offer very little chance of aggregation" (Smith 2007, p. 597).

22. In some cases, material accounts are constrained by different, physical identities (conservation of mass or energy, for example).
23. This is an oversimplification. The U.S. Bureau of Economic Analysis (U.S. BEA), for example, often relies on proxies for difficult-to-measure service outputs (such as accounting and financial services; see, for example, Griliches 1992).
24. These other species may be relevant quantity measures where other benefits are concerned, such as species existence benefits. Later in the chapter, I discuss the benefit-contingent nature of final ecological quantities in more detail.
25. For examples, see Peskin and Delos Angeles (2001), United Nations et al. (2003), Schoer (2006), and The World Bank (2006).
26. Practically speaking, there is always a trade-off in accounting between the (costly) desire for specificity and the loss of information that attends "lumpy" undifferentiated quantities.
27. This bears close resemblance to the Arrow-Debreu perspective on commodity differentiation, where the timing and location of delivered products is used to differentiate them. I thank Heal (2007) for reminding me of this analogy.
28. This is what has thwarted so-called benefit-transfer studies in environmental economics (*Ecological Economics*, Special Issue 2006).
29. The role of service zones in environmental valuation is well appreciated. For example, travel cost models require analysis of recreational substitutes, which is an inherently spatial issue.
30. In the United States, for example, the price of oil, gas, and mineral leases may not bear a close resemblance to their true social value.
31. Note that accounts like GDP are not in general composed of assets either, although assets are important to some accounts. GDP is itself built up from economic units—cars, hamburgers, haircuts—valued at the household and firm levels.
32. This reflects an overreliance on "asset" rather than "service" in the current thinking embodied in the SEEA 2003. See Section 5, 'Implications of SEEA 2003' (United Nations et al. 2003, p. 257).
33. Again, this issue is mirrored in the national accounts. We count tires sold, rather than the vehicle miles over which the tires last. Prices at the time of purchase reflect consumers' understanding of the quality difference, but ideal output measures would not aggregate products of different quality.
34. Some wetland classification systems differentiate wetlands in just this way.
35. Several NIPAs and green accounts incorporate some form of depletion analysis. Primarily, though, these efforts demonstrate the empirical difficulties of doing so. See Weber (2007) for an accounting-based view of ecological depletion.
36. Many ecologists, geographers, planners, and conservationists already think in these terms. But a consistent data network, one that helps unite disparate location-based analyses, is unavailable to these practitioners. Although the U.S. government sporadically attempts sustained monitoring at the national level, no one agency has been given such a mandate.
37. This is more than a theoretical concern. Witness, for example, recent events in Argentina. In January 2007 Argentina's government intervened in the calculation of inflation statistics by removing the official in charge and

publishing inflation estimates viewed by economists as politically rather than economically derived. Naturally, this has called the credibility of Argentina's statistical reporting into question.
38. Note, however, the ability of online communities to both provide public information goods and police their quality. Are online information communities like Wikipedia capable of producing and policing the kind of data I advocate here? It is certainly an exciting and hopeful possibility.

References

Ayres, R. and A. Kneese 1969. 'Production, Consumption, and Externalities.' *American Economic Review* LIX: 282–297.

Banzhaf, S. and J. Boyd 2005. 'The Architecture and Measurement of an Ecosystem Services Index.' RFF Discussion Paper 05–22. Washington, DC: Resources for the Future (RFF).

Boyd, J. 2006. 'The Nonmarket Benefits of Nature: What Should Be Counted in Green GDP?' *Ecological Economics* 61: 716–723.

Boyd, J. and S. Banzhaf 2007. 'What Are Ecosystem Services?' *Ecological Economics* 63(2–3): 616–626.

Binning, C., S. Cork, R. Parry, and D. Shelton 2001. *Natural Assets: An Inventory of Ecosystem Goods and Services in the Goulburn Broken Catchment.* Report of the Ecosystem Services Project. Canberra, Australia: Commonwealth Scientific and Industrial Research Organisation (CSIRO) Sustainable Ecosystems.

Carson, C. 1975. 'The History of the United States National Income and Product Accounts.' *Review of Income and Wealth* 21: 153–181.

Daily, G. 1997. *Nature's Services: Societal Dependence on Natural Ecosystems.* Washington, DC: Island Press.

Ecological Economics 2006. 'Special Issue: Environmental Benefits Transfer: Methods, Applications, and New Directions.' *Ecological Economics* 60: 2.

El Sarafy, S. 1989. 'The Proper Calculation of Income for Depletable Natural Resources.' Pages 10–18 in *Environmental Accounting for Sustainable Development.* Y.J. Ahmad, S. El Serafy, and E. Lutz. (eds), Washington, DC: The World Bank.

Griliches, Z. (ed.), 1992. *Output Measurement in the Service Sectors.* Chicago, IL: University of Chicago Press.

Heal, G. 2000. *Nature and the Marketplace: Capturing the Value of Ecosystem Services.* Washington, DC: Island Press.

——— 2007. 'Environmental Accounting for Ecosystems.' *Ecological Economics* 61: 693–694.

Heal, G. and B. Kristrom 2005. Chapter 22 in *National Income and the Environment: Handbook of Environmental Economics.* Vol. 3. Elsevier: Amsterdam, The Netherlands.

Hecht, J.E. 2000. *Lessons Learned From Environmental Accounting: Findings from Nine Case Studies.* Washington, DC: The World Conservation Union (IUCN).

——— 2005. *National Environmental Accounting: Bridging the Gap between Ecology and Economy.* Washington, DC: Resources for the Future (RFF).

Kopp, R.J. and V.K. Smith (eds), 1993. *Valuing Natural Assets.* Washington, DC: Resources for the Future (RFF).

Kremen, C. 2005. 'Managing Ecosystem Services: What Do We Need to Know about Their Ecology?' *Ecology Letters* 8: 468–479.

Mäler, K.-G. 1991. 'National Accounts and Environmental Resources.' *Environmental and Resource Economics* 1(1): 1–15.

Millennium Ecosystem Assessment 2005. *Ecosystems and Human Well-Being: Synthesis.* Washington, DC: Island Press.

Nordhaus, W.D. and E.C. Kokkelenberg (eds), 1999. *Nature's Numbers: Expanding the National Economic Accounts to Include the Environment.* Washington, DC: National Academy Press.

Olson, M. 1965. *The Logic of Collective Action: Public Goods and the Theory of Groups.* Cambridge, MA: Harvard University Press.

Peskin, H.M. 1989. 'Accounting for Natural Resource Depletion in Developing Countries.' Environmental Department Working Paper No. 13. Washington, DC: The World Bank.

Peskin, H.M. and M.S. Delos Angeles 2001. 'Accounting for Environmental Services: Contrasting the SESA and the ENRAP Approaches.' *Review of Income and Wealth* 47(2): 203–219.

Repetto, R., W. Magrath, M. Wells, C. Beer, and F. Rossini 1989. *Wasting Assets: Natural Resources in the National Income Accounts.* New York: World Resources Institute.

Schoer, K. 2006. *The German System of Environmental-Economic Accounting: Concept, Current State, and Application.* Wiesbaden: German Federal Statistical Office.

Smith, R. 2007. 'Development of the SEAA 2003 and Its Implementation.' *Ecological Economics* 61: 592–599.

——— 2006. 'Framework of Green National Accounting System in China.' Washington, DC: The World Bank.

United Nations 1984. *Integrated Environmental and Economic Accounting: An Operational Manual.* Series F, Vol. 78. New York: United Nations Statistics Division.

United Nations, European Commission, International Monetary Fund, Organisation for Economic Co-operation and Development, and The World Bank 2003. *Handbook of National Accounting, Integrated Environmental and Economic Accounting.* New York: United Nations.

U.S. BEA (U.S. Bureau of Economic Analysis) 1982. *Measuring Nonmarket Activity.* U.S. BEA Working Paper No. 2. Washington, DC: U.S. BEA.

Weber, J.-L. 2007. 'Implementation of Land and Ecosystem Accounts at the European Environment Agency.' *Ecological Economics* 61: 695–707.

6
The Challenge of Crafting Rules to Change Open-Access Resources into Managed Resources*

Elinor Ostrom

1 Garrett Hardin's model is correct but highly limited in applicability

Garrett Hardin's (1968) "tragedy of the commons" is one of the most cited articles in environment science and is assigned repeatedly to undergraduate students in Environmental Science curricula. Whenever scholars and policy discuss the problems of overuse and degradation of natural resources—whether they be fisheries, forests, irrigation systems, or the atmosphere—Hardin's article is apt to be relied upon heavily. Why has this almost metaphoric article captured so much attention? First of all, Hardin presents an extraordinarily clear and vivid picture of a pasture "open to all." Second, his assumptions about the motivation of resource harvesters are consistent with the assumptions about market participants that have proved powerful in deriving propositions regarding highly competitive markets. Viewing resource users as trapped in a tragedy of their own making is consistent with many textbooks on resource economics and the predictions derived from non-cooperative game theory for finitely repeated dilemmas (E. Ostrom, Gardner, and Walker 1994). External authorities are presumably needed to impose rules and regulations on local users since they will not do this themselves. The "scientific management of natural resources" that is frequently taught to future regulators of natural resources presents fisheries, forests, and water resources as relatively homogeneous units that are closely interrelated across a vast domain. Fish and wildlife species are

---

* Sections of this chapter draw on Elinor Ostrom, 'Coping with Tragedies of the Commons', *Annual Review of Political Science*, vol. 2 (1999), 493–535.

presented as if they always migrate over a large range. Irrigation systems are interlinked along watersheds of major river systems. This approach, as it has been applied to fisheries management, is described by Acheson, Wilson, and Steneck (1998: 391–392):

> For those trained in scientific management, it is also an anathema to manage a species over only part of its range. From the view of fisheries scientists and administrators, it is not rational to protect a species in one zone only to have it migrate into another area where it can be taken by other people due to a difference in regulations. As a result, the units to be managed range along hundreds of miles of coast and can only be managed by central governments with jurisdiction over the entire area. Lobsters, for example, extend from Newfoundland to the Carolinas; swordfish migrate from the Caribbean to Newfoundland and Iceland. From the point of view of the National Marine Fisheries Service, it makes sense to have a set of uniform regulations for the entire US coast rather than one for each state.

The belief in the capability of government analysts to design optimal rules to govern and manage common-pool resources for a large domain is shared by many academics. When common-pool resources are viewed as having a homogeneous structure and as being interlinked, simple models are developed for how they work. It is then presumed that officials—acting in the public interest—are capable of devising uniform and effective rules for an entire region. The textbooks indicate that all that is needed is reliable, statistical information on key variables for an entire region—ignoring the huge variance that may be hidden in even reliable data. Then, it is frequently presumed that one can determine the optimal harvesting level, divide this harvesting level into quotas, assign quotas to users, and allow them to buy and sell these transferable quotas. Prescriptions calling for central governments to impose uniform regulations over most natural resources are thus consistent with important bodies of theoretical work. Groups who have actually organized themselves to govern resources are frequently invisible to those who cannot imagine organization without rules and regulations issued by a central authority (see, for example, Lansing 1991; Lansing and Kremer 1994).

Extensive research does not support uniform prescriptions for an entire region to be imposed by external authorities. Hayes and Ostrom (2005) analyzed data from 163 forests located in 12 countries, of which 76 were government-owned forests that were legally designated as protected forests and 87 were public-, private-, and community-owned forested lands used for diverse purposes (see also Hayes 2006). No statistical difference was found between the vegetation densities related to officially designated, government-owned protected areas as contrasted with other property regimes. E. Ostrom and Nagendra (2006) and Gibson, Williams, and Ostrom (2005) present evidence that whether rules are monitored by users is more important to achieve the sustainability of forest resources than the formal ownership status. A large number of field studies have found that local groups of resource users have crafted a diversity of institutional arrangements for coping with common-pool resources where they have not been prevented from doing so by central authorities (McCay and Acheson 1987; Fortmann and Bruce 1988; Berkes 1989, 2007; Blomquist 1992; Bromley et al. 1992; Tang 1992; Netting 1993; Lam 1998; Meinzen-Dick 2007). These empirical studies document successful self-organized resource governance systems in diverse sectors in all parts of the world as well as cases where self-organized systems have not been successful.

We can now firmly conclude in light of extensive empirical evidence that overuse and destruction of common-pool resources is not a determinant and inescapable outcome when multiple users face a commons dilemma. Some of the key conditions of a resource, and of the users of a resource, have been identified that are conducive to local users self-organizing to find solutions to commons dilemmas (see Baland and Platteau 1996; E. Ostrom 2001). The broad design principles that characterize robust self-organized resource governance systems that have resolved commons dilemmas for long periods of time have been identified (E. Ostrom 1990, 2005) and found basically sound by other scholars (Guillet 1992a, 1992b; Morrow and Hull 1996; Bardhan 2000; Weinstein 2000; Trawick 2001; Gupta and Tiwari 2002).

A disjuncture exists between currently accepted theoretical and resultant policy recommendations related to commons and evidence from the field (Berkes et al. 1989) and the lab (E. Ostrom, Gardner, and Walker 1994). Empirical findings challenge two of the most important theoretical foundations of contemporary analysis. One foundation is the model of the human actor that is used. Resource users are explicitly thought of as norm-free, short-term, maximizers of immediate gains who will not cooperate, unless coerced by external authorities, to overcome the perverse incentives of social dilemmas in order to increase their own and others' long-term benefits. Inconsistently, government officials are depicted as capable of seeking the more general public interest and analyzing long-term patterns so as to design optimal policies. A second foundational belief for many policy analysts is that it is relatively simple to design rules to change the incentives of participants. Analysts view most resources in a particular sector as relatively similar and sufficiently inter-related that they need to be governed by the same set of rules.

In this chapter, I propose to show that these foundational assumptions are wrong and that they are a poor foundation for public policy recommendations. To do this, I will first need to define what is meant by a common-pool resource. To address the adequacy of the model of the human actor used, I then summarize the findings from a series of carefully controlled laboratory experiments of appropriation dilemmas. Given that predictions based on the model of a norm-free, myopic, and maximizing individual are not supported, except when individuals act anonymously and cannot discuss their joint problem, I then discuss the presentation of a closely related but alternative conception of human behavior—applicable to resource users and government officials alike. Humans are viewed as fallible, boundedly rational, and norm using. In complex settings, no one is able to do a complete analysis before actions are taken, but individuals learn from mistakes and are able to craft tools—including rules—to improve the structure of the repetitive situations they face.

Then, I explore the complexity of using rules as tools to change the structure of commons dilemmas. First, I describe the seven clusters of rules that affect the components of any action situation, and then describe the specific rules that are used in field settings by resource users and government agencies. An examination of the types of rules used in the field yields several important findings. First, the *number* of rules actually used in field settings is far greater than generally recognized. Second, the *type* of rules is also different. Boundary rules tend to include as co-appropriators of a resource those who are more likely to be trustworthy because they live permanently nearby and have a long-term stake in keeping a resource sustainable. Choice rules define rights and duties that are easy to understand, directly related to sustaining the biophysical structure of the research, and easy to monitor and enforce. Some rules recommended in the policy literature are not found among the rules used by self-organized systems.

Given the complexity of the process of designing rules to regulate the use of common-pool resources, I argue that all policy proposals must be considered as experiments. No one can possibly know whether a proposed change in rules is among the more optimal rule changes or even whether a rule change will lead to an improvement. All policy experiments have a positive probability of failing. Then I discuss how the parallel efforts by a large number of local resource users to search out and find local rule configurations may find better rule combinations over the long term while top-down design processes are more limited in their capacities to search and find appropriate rules. All forms of decision making have limits. Thus, we need to understand the limits of fully decentralized, independent resource governance systems and the importance of building polycentric governance systems with considerable overlap to combine the strengths of parallel search and design processes with the strengths of larger systems in conflict resolution, acquisition of scientific knowledge, monitoring the performance of local systems, and the regulation of common-pool resources that are more global in their scope. The resulting polycentric governance systems are also not directed by a single center bit rather a form of complex adaptive systems.

2   What is a common-pool resource?

A common-pool resource is a natural or man-made resource from which it is difficult to exclude or limit users once the resource is provided by nature or produced by humans (E. Ostrom, Gardner, and Walker 1994). What one person consumes removes resource units from what is available to others. Thus, the trees or fish harvested by one user are no longer available for others. Commons share the difficulty of excluding beneficiaries with public goods, while the subtractability of the resource units is shared with private goods. In order to provide a focus, I will primarily examine renewable natural resources as exemplars of common-pool resources, but the theoretical arguments are relevant to man-made common-pool resources as well.

When the resource units (e.g., the fish, trees, or water) produced by a common-pool resource have a high value and institutional constraints do not restrict the way these units are appropriated, individuals face strong incentives to appropriate more and more resource units leading eventually to congestion, overuse, and even the destruction of the resource itself. Because of the difficulty of excluding beneficiaries, the free-rider problem is a potential threat to any efforts to reduce harvesting and improve the long-term outcomes achieved from the use of the common-pool resources. If some individuals are cooperative and do not harvest as many units, the benefits so generated are shared with others whether the others cut back on their harvesting or not. Some individuals are likely to free ride on the costly actions of others unless ways are found to reduce free riding as an attractive strategy. Once some users free ride, others are likely to follow suit and overharvesting may soon be the outcome.

Consequently, one of the important problems facing the joint users of a common-pool resource is known as the "appropriation problem" given the potential incentives in all jointly used common-pool resources for individuals to appropriate more resource units when acting independently than they would if they could find some way of coordinating their appropriation activities. Joint users of a common-pool resource often face many other problems including assignment problems, technological externality problems, provision problems, and maintenance problems (E. Ostrom, Gardner, and Walker 1994; E. Ostrom and Walker 1997). And, the specific character of each of these problems differs substantially from one resource to the next. Here, I will focus primarily on appropriation problems since they are what most analysts associate with "the tragedy of the commons."

3   A baseline appropriation situation

Let us start with a static, "institution-free," baseline situation that is as simple as feasible without losing crucial aspects of the problems that real appropriators face in the field. This will let us understand the outcomes predicted and achieved in such a baseline situation and the processes involved in changing the structure by changing rules affecting it. This institution-free, static, baseline situation is composed of the following:

1. A set of n symmetric appropriators who are interested in withdrawing resource units from a common-pool resource.
2. No differentiation exists in the positions these appropriators hold relevant to the common-pool resource. In other words, there is only one position of appropriator.
3. Appropriators must decide how to allocate their time and effort in each time period. We can think of these appropriators as being "endowed" with a set of assets, e, that they are free to allocate during each time period to two activities. Appropriators must decide, for example, for each time period between spending time trying to harvest resource units from the common-pool resource and using time in their next best opportunity, such as working in a local factory. To simplify the problem, let us assume that all appropriators have the same endowment, face the same labor market, and can earn a fixed wage for any time they allocate to working for a factory.

4. The actions they take affect the amount of resource units that can be appropriated from the common-pool resource or wages earned in the labor market.

5. Transformation functions map the actions of all of the appropriators given the biophysical structure of the resource itself onto outcomes. While these functions are frequently stochastic in field settings and affected by many variables in addition to the actions of individuals, let us assume here determinant functions. The wage function simply multiplies the amount of time allocated to it by whatever is the standard wage. The appropriation function is a concave function, F, which depends on the number of assets, x_i, allocated to appropriation from the common-pool resource. Initially, the sum of individuals' actions, $\sum x_i$, generates better outcomes than the safe investment in wage labor. If the appropriators decide to allocate a sufficiently large number of their available assets, the outcome they receive is less than their best alternative. Such a function is specified in many resource economics textbooks based on Gordon (1954) and Scott (1955).

6. Regarding information, let us assume that appropriators know the shape of the transformation function and know that they are symmetric in assets and opportunities. Information about outcomes is generated after each decision round is completed.

7. Payoff rules specify the value of the wage rate and the value of the resource units obtained from the common-pool resource. As analysis in E. Ostrom, Gardner, and Walker (1994), the payoff to an appropriator is given by:

we if $x_i = 0$

$$w(e - x_i) + \left(x_i / \sum x_i\right) F\left(\sum x_i\right) \quad \text{if } x_i > 0. \tag{1}$$

Basically, if appropriators put all of the assets into the available wage labor, they receive a known return equal to the amount of their endowment times the wage rate. If appropriators put some of their assets into wage labor and some into the common-pool resource, they get part of their return from wages and the rest from their proportional investment in the common-pool resources times the total output of the common-pool resource as determined by function F.

3.1 Assumptions about actors

To explain and predict the outcome of any situation, one needs to specify four key characteristics about the actors who are participating in the situation: (1) the type of preferences held, (2) how information is processed, (3) the formula or heuristic used for making decisions, and (4) the resources brought to the situation. The theory of complete rationality uses the assumptions that (1) individuals have a complete and transitive ordering of preferences over all outcomes that is monotonically related to only their own returns, (2) all relevant information generated by the situation is used in making decisions, (3) actors maximize their own expected returns, and (4) all needed resources to act in this situation are possessed. The theory of norm-free, complete rationality has proved to be extremely useful in a diversity of circumstances where the institutional arrangements reduce the number of options and complexity of the situation and reward those who maximize expected returns to self and punish those who do not. When such situations are completely specified, clear predictions of equilibrium outcomes can be derived. Behavior in experimental laboratories and in the field closely approximates the predicted equilibrium in simple action situations where selection pressures retain those who maximize their own expected returns and thin out those who do not.

The theory of norm-free, complete rationality is also useful in a variety of other situations to enable the analyst to undertake a full analysis and predict equilibrium outcomes. If behavior deviates from the predicted outcomes, one has a clear benchmark for knowing how far behavior deviates from that predicted by this theory. We will thus initially use the theory of norm-free, complete rationality and the theory of finitely repeated games to predict what the outcome would be if a set of experimental subjects were to face a fully specified baseline appropriation situation as outlined above. We will later modify this set of assumptions in light of the evidence obtained in the experimental laboratory (and supplemented by field studies).

3.2   Predicted outcomes for a common-pool resource in the laboratory

Laboratory experiments provide an opportunity to observe how humans behave in situations that are very simple when compared to field settings, but nonetheless, characterize essential common elements of relevant field situations. In the laboratory experiments conducted at Indiana University, we thought it crucial to examine behavior in an appropriation situation with a nonlinear transformation function and a sufficient number of players that knowledge of outcomes did not automatically provide information about each player's actions. In this chapter, I can only briefly discuss the results of these experiments. All procedures and specifications are thoroughly documented in E. Ostrom, Gardner, and Walker (1994) and in journal articles cited therein. In the baseline experiments, we utilized the following equation for the transformation function, F.

$$23\left(\sum x_i\right) - 25\left(\sum x_i\right)^2 \tag{2}$$

Eight subjects participated in all experiments discussed in this chapter and each subject was assigned 25 tokens as their endowment in each round of play. Their outside opportunity was valued at \$0.05 per token. They earned \$0.01 on each outcome unit they received from investing tokens in the common-pool resource. Subjects were informed that they would participate in an experiment that would last no more than 2 hours, but the number of rounds in each experiment varied between 20 and 30 rounds. The situation was described as involving a choice between investing in either of two markets having the structure as specified above. In addition to being told the payoff function specifically, subjects were provided with look-up tables that eased their task of determining outcomes depending on their own and others' decisions. All experiments reported on in this chapter involved subjects who had prior experience in similar experiments.

With these specifications, the predicted outcome for a finitely repeated game where subjects are not discounting the future is for each subject to invest 8 tokens in the common-pool resource for a total of 64 tokens (the Nash Equilibrium). The players could, however, earn more if the total number of tokens invested was 36 tokens, rather than 64 tokens, in the common-pool resource. This optimal level of investment would earn each subject \$0.83 per round. The baseline experiment is a clear example of a commons dilemma.

3.3   Outcomes of a *N*-person repeated appropriations dilemma

As documented in E. Ostrom, Gardner, and Walker (1994), subjects interacting in baseline experiments substantially overinvested as predicted. On average, subjects received −3 percent of optimum (E. Ostrom, Gardner, and Walker 1994: 116). However, at the individual level, subjects rarely invested 8 tokens, which is the predicted Nash Equilibrium. Instead, there was an unpredicted and strong pulsing pattern in all experiments. Individuals appear to increase their investments in the common-pool resource until there is a strong reduction in yield, at which time they tend to reduce their investments leading to an increase in yields. The pattern is repeated over time. No game-theoretical explanation exists for the pulsing pattern.

Subjects explained that they were using several rules of thumb or heuristics in response to postexperiment questioning. One of the heuristics was to invest more in the common-pool resource whenever the rate of return on the previous round was above $0.05 (what they could earn in their next best alternative) and less if the return was below $0.05. Equilibrium is really never reached at the individual level. Thus, "each player is continually having to revise his or her response to the current 'anticipated' situation. This strategic turbulence on top of an already complex task increases the chances that a player may not attempt a best-response approach to the task but rather invoke simple rules of thumb..." (E. Ostrom, Gardner, and Walker 1994: 121–122).

These laboratory experiments have been replicated by other researchers (Rocco and Warglein 1995; Cardenas 2000; Cardenas, Stranlund, and Willis 2000; Casari and Plott 2003) with similar results.

3.4   Structural changes in the lab

In addition to the baseline experiments, we have explored how changes in the rules affect outcomes. In the lab rule changes are operationalized by the set of instructions given to subjects and in the procedures adopted within the experiment. The first structural change we used is an information rule change. Instead of forbidding all communication among subjects, as in the baseline experiments, subjects were now authorized to communicate with one another in a group setting before returning to their terminals to make their own private decisions. This rule change gave subjects an opportunity for "cheap talk." In cheap talk conditions, agreements made by subjects are not enforced by an external authority. Cheap talk is viewed within the context of noncooperative game theory as irrelevant. The same outcome is predicted as in the baseline experiment.

In a second series of experiments, we changed the authority and payoff rules to allow subjects to sanction one another at a cost to themselves. Using this rule change enables subjects to produce a benefit for all at a cost to themselves. The game-theoretic prediction is that no one will choose the costly sanctioning option. Third, we changed the authority rule to allow subjects to covenant with one another to determine their investment levels and to adopt a sanctioning system if they wished. Again, the predicted outcome is the same. In all three of these changed appropriation experiments, however, subjects demonstrate their willingness and ability to search out and adopt better outcomes than those predicted.

3.4.1   Face-to-face communication

In the repeated communication experiments, subjects made ten rounds of decisions in the context of the baseline appropriation game. An announcement then told them they would have an open group discussion before each of the continuing rounds of the experiment. The subjects left their terminals and sat in a group facing one another. After each discussion, they returned to their terminals and entered their anonymous decisions. Subjects used face-to-face communication to discuss together what strategy would gain them the best outcomes and to agree on what everyone should invest in the subsequent rounds. After each decision round, they were informed what their aggregate investments had been, but *not* the decisions of individual players. Thus, they learned whether total investments were greater than their agreement. While in many rounds, subjects kept their promises, some defections did occur. If promises were not kept, subjects used this information to castigate the unknown participant who had not kept to their agreement.

Subjects in the 25-token baseline experiments had received total returns that were slightly below zero, while in the communication experiments, they obtained on average 62 percent of the maximum available returns (with variation across experiments). The defection rate was 13 percent. Our conclusion in completing an analysis of these experiments was as follows:

> Communication discussions went well beyond discovering what investments would generate maximum yields. A striking aspect of the discussion rounds was how rapidly subjects, who had not had an opportunity to establish a well-defined community with strong internal norms, were able to devise their own agreements and verbal punishments for those who broke those agreements.... In many cases, statements like "some scumbucket is investing more than we agreed upon" were a sufficient reproach to change defectors' behavior.
>
> (E. Ostrom, Gardner, and Walker 1994: 160)

The process of internalizing norms regarding the importance of keeping promises is evidenced by several of their behaviors. Simply promising to cut back on their investments in the common-pool resource led most subjects to change their investment pattern. Second, subjects were indignant about evidence of investment levels higher than that promised and expressed their anger openly. Third, those who broke their promise tended to revert to the promised level after hearing the verbal tongue-lashing of their colleagues.

3.4.2 Sanctioning experiments

Participants in many smaller common-pool resources in the field are usually able to communicate with one another on a face-to-face basis either in formally constituted meetings or at social gatherings. In most field settings, however, participants also devised a variety of formal or informal ways of sanctioning one another if rules are broken. Engaging in costly monitoring and sanctioning behavior is, however, not consistent with the theory of norm-free, complete rationality (Elster 1989: 40–41). Thus, it was important to ascertain whether subjects in a controlled setting would actually pay in order to assess a financial punishment on the behavior of other participants. The short answer to this question is yes.

In all of the sanctioning experiments, subjects played ten rounds of the baseline game modified so that the individual contributions in each round were reported as well as the total outcomes. Subjects were then told that in the subsequent rounds they would have an opportunity to pay a fee in order to impose a fine on the payoffs received by another player. The fees ranged in diverse experiments from $0.05 to $0.20 and the fines from $0.10 to $0.80. Much more sanctioning occurred in these experiments than the zero level predicted. Subjects reacted both to the cost of sanctioning and to the fee/fine relationships. They sanctioned more when the cost of sanctioning was less and when the ratio of the fine to the fee was higher.

Sanctioning was primarily directed at those who invested more in the common pool resource, but a few sanctions appear to be directed by those who had been fined in a form of "blind revenge" against those whose investments were lower than others and were thus suspected of having sanctioned them. Since we first report the results of our sanctioning experiments (E. Ostrom, Walker, and Gardner 1992), many other scholars have designed experiments where subjects were given the option of sanctioning others and have consistently found higher levels of sanctioning than predicted by theory (Fehr and Gächter 2000; Carpenter, Matthews, and Ong'ong'a 2004; Anderson and Putterman 2005).

In this set of experiments, subjects were able to increase their returns modestly to 39 percent of maximum, but when the costs of fees and fines were subtracted from the total, these gains are wiped out. When subjects were given a single opportunity to communicate prior to the implementation of sanctioning capabilities, they were able to gain an average of 85 percent of the maximum payoffs (69 percent when the costs of the fees and fines were subtracted).

3.4.3   Covenanting experiments

In self-organized field settings, the opportunity to punish those who free ride is much more likely to emerge from an endogenous process of crafting their own rules, including the punishments that should be imposed if these rules are broken. Spending time and effort designing rules creates a public good for all of those involved and is thus a second-level dilemma. Noncooperative game theory predicts that participants will not undertake such efforts. This is the theoretical foundation for the policy advice that rules must be imposed on participants by external authorities who then assume responsibility for monitoring and enforcing these rules. Since self-organized rules are found in many local common-pool resource situations, it does appear that participants frequently do design their own rules contrary to the theoretical prediction. Few scholars are able to witness these processes, however.

Subjects experienced with baseline and sanctioning experiments were recalled and given an opportunity to have a "constitutional convention" in the laboratory. They could decide whether or not they would like to have access to a sanctioning mechanism like the one described above, how much the fines and fees should be, and on the joint investment strategy that they would like to adopt. Four out of six experimental groups adopted an agreement and specified the number of tokens they would invest and the level of fines to be imposed. The fines determined by the participants ranged in size from $0.10 to $1.00. The groups that crafted their own agreements were able to achieve an average of 93 percent of the maximum in the periods after their agreement. And, the defection rate for these experiments was only 4 percent. The two groups that did not agree to their own covenant did not fare as well. They averaged 56 percent of the maximum available returns and faced a defection rate of 42 percent. Consequently, those subjects who used an opportunity to covenant with one another to agree on a joint strategy chose their own level of fines, and received very close to optimal results based entirely on their own promises and their own willingness to monitor and sanction one another when it was occasionally necessary (see Frohlich, Oppenheimer, and Eavey 1987 for similar findings).

4   Developing a theory of human behavior consistent with evidence from the lab

The appropriation experiments briefly summarized above provide the following picture of behavior in *N*-person, finitely repeated, commons-dilemma situations:

1. When individuals are held apart and unable to communicate on a face-to-face basis, they overuse a common-pool resource.
2. Individuals initially use opportunities for face-to-face discussions to share their understanding of how their actions affect the joint outcomes and arrive at a common understanding of the best joint strategy available to them.
3. Individuals tend to use heuristics in dealing with complex problems and these vary in their capabilities to cope with changing configurations of actions by other participants.
4. Individuals are willing to promise others, whom they assess as being trustworthy, that they will adopt a joint plan of action. Most individuals keep their promises (even in situations where substantial advantage can accrue for breaking the promise).
5. If agreements are broken, individuals become indignant and use verbal chastisements when available. They are also willing to use (and overuse) costly sanctions, but they do not use grim trigger strategies.
6. When given an opportunity to craft their own rules and sanction nonconformance to these rules, many (but not all) groups are willing to do so and then tend to achieve close to optimal results.

In other words, individuals tend to rely on diverse heuristics in response to complexity. Without communication and agreements on joint strategies, overuse of a common-pool resource is highly likely. On the other hand, individuals are willing to discuss ways of increasing their own and others' payoffs over time. Many are willing to make contingent promises when others are assessed as trustworthy. A substantial number of individuals, but not all, are trustworthy and reciprocate the trust that has been extended. When noncooperative behavior is discovered, individuals are willing to use retribution in a variety of forms.

Assuming that individuals have the capability to engage in problem solving to increase long-term payoffs, to make promises, to build reputations for trustworthiness, to reciprocate trustworthiness with trust, and to punish those who are not trustworthy, leads to a different policy conclusion than assuming that individuals seek their own short-term, narrow interests even when presented with repeated situations where everyone's joint returns could be substantially increased. Using the latter theory leads to the policy advice that rules to reduce overuse must be devised by external authorities and enforceably imposed on local users. This was the foundation for most policy prescriptions regarding the regulation of common-pool resources during the second half of the last century.

A better foundation is to assume that humans may not be able to analyze all situations fully, but that they will make an effort to solve complex problems by crafting regularized procedures and will be able to draw on inherited capabilities to learn norms of behavior, particularly reciprocity (Bendor 1987; E. Ostrom 1998). A behavioral theory of boundedly rational and norm-using behavior views all policies as experiments and asks what processes of search and problem solving are more likely to arrive at better experiments. The key problems to be solved are how to ensure that those using a common-pool resource share a similar and relatively accurate view of the problems they need to solve, how to devise rules to which most can contingently agree (Levi 1988), and how to monitor activities sufficiently so that those who break agreements through error or succumbing to the continued temptations that exist in all such situations are sanctioned, and thus trust and reciprocity are supported rather than undermined (Bendor and Mookherjee 1990).

Common-pool dilemmas never fully disappear even in the best operating systems. The temptation to cheat always exists. No amount of monitoring and sanctioning reduces the temptation to cheat entirely. Instead of thinking of overcoming or conquering tragedies of the commons, effective governance systems cope better than others with the ongoing need to encourage high levels of trust at the same time as needing to monitor actions and sanction rule infractions.

Presenting this difference in a theoretical perspective based on carefully designed laboratory experiments is the first task that I set out to accomplish. Boundedly rational, local users are potentially capable of changing their own rules, enforcing the rules they agree upon, and learning from experience to design better rules. The next task is to show why multiple, boundedly rational, local users are better at designing rules than a team of boundedly rational officials in a central agency. To do this, we need to draw on research about the type of rules used in the field.

5 Experimenting with rules in the field

With this change in perspective, we can think of appropriators trying to understand the biophysical structure of a resource they are using and how to affect each other's incentives so as to increase the probability of sustainable and more efficient use over the long term. Instead of being given a set of instructions with the transformation function fully specified—as subjects do in a lab experiment—appropriators in the field have to explore and discover the biophysical structure of a particular resource that will differ on key parameters from similar resources in the same region. Further, they have to cope with considerable uncertainty related to the weather, complicated growth patterns of biological systems that may at times be chaotic in nature, and external price fluctuations affecting the costs of inputs and value of outcomes (see Wilson et al. 1991, 1994; Wilson, Yan, and Wilson 2007). In addition to the physical changes that they can make in the resource, the tools they can use to change the structure of the action situations they face consist of seven clusters of rules that directly affect the components of their own action situations. Specifically, the rules they can change affect the working parts of an action situation or a game (E. Ostrom 2005). They include the following:

> Boundary rules affect the characteristics of the *participants*.
> Position rules differentially affect the capabilities and responsibilities of those in *positions*.
> Choice rules affect the *actions* that participants in positions may, must, or must not do.
> Scope rules affect the *outcomes* that are allowed, mandated, or forbidden.
> Aggregation rules affect how individual actions are *transformed* into final outcomes.
> Information rules affect the kind of *information* present or absent in a situation.
> Payoff rules affect assigned *costs and benefits* to actions and outcomes.

Given the nonlinearity and complexity of action situations in the field, it is rarely easy to predict what effect a change in a particular rule will produce. For example, a change in a boundary rule to restrict the entry of other potential users simultaneously reduces the number of individuals who are tempted to break rules, but it also reduces the number of individuals who monitor what is happening or contribute funds toward hiring a guard. Thus, the opportunities for rule breaking may increase. Further, the cost of a rule infraction will be spread over a smaller group of appropriators and, thus, the harm to any individual may be greater. Assessing the overall effects of a change in boundary rules is a non-trivial analytical task (for examples, see Weissing and Ostrom 1991a, 1991b; Acheson and Gardner 2005). Instead of conducting such a complete analysis, appropriators are more apt to use their past experience in using the resource and with one another to experiment with different rule changes until they find a combination that seems to work in their setting.

To understand the types of tools that appropriators from common-pool resources use somewhat better, let us examine in some detail the kind of boundary, choice, payoff, and position rules found in field settings. These four clusters of rules are the major tools we have repeatedly found that affect the performance of common-pool resource systems. Information, scope, and aggregation rules are utilized to complement changes induced by these four rules.

For the several decades, colleagues at or associated with the Workshop in Political Theory and Policy Analysis at Indiana University have studied a very large number of irrigation systems, forests, inshore fisheries, and groundwater basins, as well as other common-pool resources (see Schlager 1990; Tang 1992; Schlager, Blomquist, and Tang 1994; Lam 1998; Gibson, McKean, and Ostrom 2000; Gautam and Shivakoti 2005; Nagendra, Karna, and Karmacharya 2005). We have collected an immense archive of original case studies conducted by many different scholars on all sectors in all parts of the world (see the Digital Library of the Commons for extensive citations, http://dlc.dlib.indiana.edu/). Using the Institutional Analysis and Development (IAD) framework, multiple research teams have developed structured coding forms to help identify the specific kinds of action situations faced in the field as well as the types of rules that users have evolved over time to try to govern and manage their resource effectively.

5.1 Using boundary rules

Policy analysts frequently recommend limiting the number of persons allowed to appropriate from a common-pool resource so that the level of appropriation is reduced or to require users to obtain a license before harvesting. Boundary rules affect the types of participants with whom other participants will interact. If contingent cooperation is perceived to be a possibility, then an important way to enhance the likelihood of using reciprocity norms is to increase the proportion of participants who are well known in a community, have a long-term stake in that community, and would try to build their reputation for trustworthiness in the community. Reducing the number of users but opening the resource to strangers willing to pay a license fee, who lack a long-term interest in the sustainability of a particular resource, may reduce the level of trust and willingness to use reciprocity and thus increase enforcement costs substantially.

As shown in Table 6.1, we identified 27 boundary rules described by case-study authors as having been used in at least one common-pool resource somewhere in the world (E. Ostrom, Gardner, and Walker 1994). While some systems use only a single boundary rule, many use two or three of these rules in combination. Boundary rules can be broadly classified in three general groups defining how individuals gain authority to enter and appropriate resource units from a common-pool resource. The first type of boundary rule relates to an individual's citizenship, residency, or membership in a particular organization. Forestry

Table 6.1 Attributes used in boundary rules to define who is authorized to appropriate from a common-pool resource

Attributes		*Conditions*
Residency or membership	**Personal characteristics**	**Relationship with resource**
National	Ascribed	Continued use of resource
Regional	Age	Use of specified technology
Local community	Caste	Long-term rights based on:
Organization	Clan	Ownership of a proportion of
(e.g., co-op)	Class	annual flow of resource units
	Ethnicity	Ownership of land
	Gender	Ownership of nonland asset
	Race	(e.g., berth)
	Acquired	Ownership of shares in a private
	Education level	firm
	Skill test	Ownership of a share of the
		resource system
		Temporary use-rights acquired through:
		Auction
		Licenses
		Lottery
		Per-use fee
		Registration
		Seasonal fees

Source: Adapted from E. Ostrom (2005: 224).

and fishing user groups frequently require members to have been born in a particular location.

A second broad group of rules relates to individual ascribed or acquired personal characteristics. Other user groups may require that appropriation depends on ethnicity, clan, or caste. A third group of boundary rules relates to the relationship of an individual with the resource itself. Using a particular technology or acquiring appropriation rights through an auction or a lottery are examples of this type of rule. About half of the rules relate to the characteristics of the users themselves. The other half involves diverse relationships with the resource.

In a systematic coding of those case studies for which sufficient information existed about rules related to inshore fisheries in many parts of the world, Schlager (1994) coded 33 user groups out of the 44 groups identified as having at least some rules regarding the use of the resource. All 33 groups depended on a combination of 14 different boundary rules (Schlager 1994: 258) and none relied on a single boundary rule. Thirty out of 33 groups (91 percent) limited fishing to those individuals who lived in a nearby community, while 13 groups also required membership in a local organization. Consequently, most inshore fisheries organized by the users themselves restrict fishing to those individuals who are well known to each other, who have a relatively long-term time horizon, and who are connected to one another in multiple ways (see Taylor 1982; Singleton and Taylor 1992).

After residency, the next most frequent type of rules, used in two-thirds of the organized subgroups, involves the type of technology that a potential fisher must be willing to use. These rules are often criticized by policy analysts, since gear restrictions tend to reduce the "efficiency" of fishing. Used in combination with choice rules that assign fishers using one type of gear to one area of the fishing groups and fishers using another type of gear to a second area, however, these solve conflicts among noncompatible technologies. Many gear restrictions also place a reduced load on the fishery itself and thus help to sustain longer-term use of the resource. They also reduce the cost of monitoring conformance with rules. In addition, other groups used a wide diversity of rules shown in Table 6.1. The key finding for the argument presented in this chapter is that Schlager did not find that any particular boundary rule was correlated with higher performance levels. Schlager did find, however, that the 33 groups who had at least one boundary rule tended to be able to solve common-pool problems more effectively than the 11 groups who had not crafted boundary rules.

In a related study of 43 small- to medium-sized irrigation systems managed by farmers or by government agencies, Tang (1992) found that the variety of rules used in irrigation was smaller than among inshore fisheries. The single most frequently used boundary rule, used in 32 of the 43 systems (74 percent), was that an irrigator must own land in the service area of an irrigation system (Tang 1992: 84–85). All of the government-owned and government-operated irrigation systems relied on this rule and only this rule. Many of the user-organized systems relied on other rules or land ownership combined with other rules. Among the other rules used were ownership of a proportion of the flow of the resource, membership in a local organization, and a per-use fee. Tang (1992: 87) found a strong negative relationship between reliance on land as the *sole* boundary requirement and performance. Over 90 percent of the systems using other boundary rules or a combination of rules including land ownership were rated positively in the level of mainte-nance achieved and in the level of rule conformance, while less than 40 percent of those systems relying solely on land ownership were rated at a higher performance level (p = 0.001). Many government systems are designed on paper to serve an area larger than they are actually able to serve when in operation, because of a variety of factors including the need to show as many posited beneficiaries as possible to justify the cost of construction (see Palanisami 1982; Repetto 1986; Shivakoti and Ostrom 2002). After construction, authorized irrigators find water to be very scarce and are unwilling to abide by choice rules or contribute to the maintenance of the system.

The rich diversity of boundary rules used by appropriators in the field appears to be a way of ensuring that the appropriators will be relating to others who live nearby and have a long-term interest in sustaining the productivity of the resource. One way of coping with the commons is thus changing the composition of who uses a common-pool resource to increase the proportion of participants who have a long-term interest, who are more likely to use reciprocity, and who can be trusted. Central governments tend to use a smaller set of rules and some of these may open up a resource to strangers without a longer-term commitment to the resource.

5.2 Using choice rules

Choice rules are also a major tool used to regulate common-pool resources. Some rules involve a simple formula. Many forest resources, for example, are closed to all forms of harvesting during one portion of

Table 6.2 Types of choice rules

Allocation formula for appropriation rights	Basis for allocation formula
Percentage of total available units per period	Amount of land held
	Amount of historical use
Quantity of resource units per period	Location of appropriator
	Quantity of shares of resource owned
Location	Proportion of resource flow owned
Time slot	Purchase of periodic rights at auction
Rotational order	Rights acquired through periodic lottery
Appropriate only during open seasons	Technology used
	License issued by a governmental authority
Appropriate only resource units meeting criteria	Equal division to all appropriators
Appropriate whenever and wherever	Needs of appropriators (e.g., type of crop)
	Ascribed characteristic of appropriator
	Membership in organization
	Assessment of resource condition

Source: Adapted from E. Ostrom (2005: 229).

the year and open for extraction by all who meet the boundary rules during an open season. Most choice rules, however, have two components. In Table 6.2, the eight allocation formulas used in the field are shown in the left column. A fisher might be assigned to a fixed location (a fishing spot) or to a fixed rotational schedule, a member of the founding clan may be authorized to cut timber anywhere in a forest, while an irrigator might be assigned to a fixed percentage of the total water available during a season or to a fixed time slot. In addition to the formula used in a choice rule, most rules required a basis for the assignment. For example, a fisher might be assigned to a fixed location based on a number drawn in a lottery, on the purchase of that spot in an auction, or on the basis of his or her historical use. An irrigator might be assigned to a fixed rotation based on the amount of land owned, the amount of water used historically, or the specific location of the irrigator.

If all bases were combined with all of the formula, there would be 112 different choice rules (8 allocation formulas × 14 bases). A further complication is that the rules for one product may differ from those of another product in the *same* resource. In regard to forest resources, for example, children may be authorized to pick fruit from any tree located in a forest so long as it is for their own consumption, women may be authorized to collect so many headloads of dead wood for domestic fire-wood and certain plants for making crafts, while *shaman* are the only ones authorized to collect medicinal plants from a particular location in a forest (Fortmann and Bruce 1988). Appropriation rights to fish are frequently related to a specific species. Thus, the exact number of rules that are actually used in the field is difficult to compute since not all bases are used with all formulas, but many rules focus on specific products. A still further complication is that the rules may regularly change over the course of a year depending on resource conditions.

Schlager (1994: 259–260) found that all 33 organized subgroups used one of the five basic formulas in their choice rules. Every user group included in her study assigned fishers to fixed locations using a diversity of bases including technology, lottery, or historical use. Thus, spatial demarcations are a critical variable for inshore fisheries. Nine user groups required fishers to limit their harvest to fish that met a specific size requirement, while seven groups allocated fishers to fishing spots using a rotation system and seven other groups only allowed fishing locations to be used during a specific season. Four groups allocated fishing spots for a particular time period (a fishing day or a fishing season).

An important finding, given the puzzles addressed in this chapter, is that the authority rule most frequently recommended by policy analysts (see Anderson 1986, 1992; Copes 1986) is *not* used in any of the coastal fisheries included in Schlager's study. Thus, no attempt was made "by the fishers involved to directly regulate the quantity of fish harvested based on an estimate of the yield. This is particularly surprising given that the most frequently recommended policy prescription made by fishery economists is the use of individual transferable quotas based on estimates on the economically optimal quantity of fish to be harvested over the long run" (Schlager 1994: 397). In an independent study of 30 traditional fishery societies, James Wilson and colleagues also noted the surprising absence of quota rules:

> All of the rules and practices we found in these 30 societies regulate "how" fishing is done. That is, they limit the times fish may be caught, the locations where fishing is allowed, the technology permitted, and the stage of the life cycle during which fish may be taken. None of these societies limits the "amount" of various species that can be caught. Quotas, the single most important concept and tools of scientific management, is conspicuous by its absence.
>
> (Acheson, Wilson, and Steneck 1998: 397; see Wilson et al. 1994)

Local inshore fishers, when allowed to manage a riparian area, thus use rules that differ substantially from those recommended by advocates of scientific management. Fishers have to know a great deal about the ecology of their inshore region including spawning areas, nursery areas, the migration routes of different species, and seasonable patterns just in order to succeed as fishers. Over time, they learn how "to maintain these critical life-cycle processes with rules controlling technology, fishing locations, and fishing times. Such rules in their view are based on biological reality" (Acheson, Wilson, and Steneck 1998: 405).

In the irrigation systems studied by Tang (1992: 90–91), three types of choice rules are used most frequently: (1) a fixed time slot is assigned to each irrigator (19 out of the 37 cases for which data is available, and in 10 out of 12 government-owned systems), (2) a fixed order for a rotation system among irrigators (13 cases), and (3) a fixed percentage of the total water available during a period of time (5 cases). Three poorly performing systems with high levels of conflict use no authority rule at all. A variety of bases were used in these rules such as "amount of land held, amount of water needed to cultivate existing crops, number of shares held, location of field, or official discretion" (Tang 1994: 233). Farmers also do not use rules that assign a specific quantity of water to irrigators other than in the rare circumstances where they control substantial amounts of water in storage (see Maass and Anderson 1986). Fixed time slot rules allow farmers considerable certainty as to when they will receive water without an equivalent certainty about the quantity of water that will be available in the canal. When the order is based on a share system, simply owning land next to an irrigation system is not enough. A farmer must purchase one or more shares to irrigate for a particular time period. Fixed time allocation systems, which are frequently criticized as inefficient, do economize greatly on the amount of knowledge farmers have to have about the entire system and on monitoring costs. Spooner (1974) and Netting (1974) described long-lived irrigation systems in Iran and in Switzerland where there was perfect agreement on the order and time allotted to all farmers located on a segment of the system, but no one knew the entire sequence for the system as a whole.

Tang also found that many irrigation systems use different sets of rules depending on the availability of water. During the most abundant season, for example, irrigators may be authorized to take water whenever they need it. During a season when water is moderately available, farmers may use a rotation system where every farmer is authorized to take water for a fixed amount of time during the week based on the amount of land to be irrigated. During scarcity, the irrigation system may employ a special water distributor who is authorized to allocate water to those farmers who are growing crops authorized by the irrigation system and are most in need.

The diversity of rules devised by users greatly exceeds the limited choice rules recommended in textbook treatments of this problem. Appropriators thus cope with the commons by a wide variety of rules affecting the actions available to participants and thus their basic set of strategies. Given this wide diversity of rules, it is particularly note-worthy that rules assigning appropriators a right to a specific quantity of a resource are used so infrequently in inshore fisheries and irrigation systems. (They are used more frequently when allocating forest products where the quantity available and the quantity harvested are much easier to measure (Agrawal 1994).) To assign an appropriator a specific quantity of a resource unit requires that those making the assignment know the total available units. In water resources where there is storage of water from one season to another and reliable information about the quantity of water is available, such rules are more frequently utilized (Blomquist 1992; Schlager, Blomquist, and Tang 1994).

5.3 Using payoff and position rules

One way to reduce or redirect the appropriations made from a common-pool resource is to change payoff rules so as to add a penalty to actions that are prohibited. Many user groups also adopt norms that those who are rule breakers should be socially ostracized or shunned and individual appropriators tend to monitor each other's behavior rather intensively. Three broad types of payoff rules are used extensively in the field: (1) the imposition of a fine, (2) the loss of appropriation rights, and (3) incarceration. The severity of each of these types of sanctions can range from very low to very high and tends to start out on the low end of the scale. Inshore fisheries studied by Schlager relied heavily on shunning and other social norms and less on for-mal sanctions. Thirty-six of the 43 irrigation systems studied by Tang used one of these three rules and also relied on vigorous monitoring of each other's behavior and shunning of rule breakers. The 7 systems that did not self-consciously punish rule infractions were all rated as having poor performance. Fines were most typically used (in 21 cases) and incarceration the least (in only 2 cases). Fines tend to be gradu-ated depending on the seriousness of the infractions and the number of prior infractions. The fines used for a first or second offence tend to be very low.

Passing rules that impose costs is relatively simple. The real diffi-cult task is monitoring behavior to ascertain if rules are being broken. Self-organized fisheries tend to rely on self-monitoring more than the creation of a formal position of guard. Most inshore fishers now use short-wave radios as a routine part of their day-to-day operations allow-ing a form of instant monitoring to occur. An official of a West Coast Indian tribe reports, for example, that "it is not uncommon to hear messages such as 'Did you see so-and-so flying all that net?' over the short-wave frequency, a clear reference to a violation of specified gear limits" (cited in Singleton 1998: 134). Given that most fishers will be listening to their short-wave radio,

> such publicity is tantamount to creating a flashing neon sign over the boat of the offender. Such treatment might be preceded or followed by a direct approach to the rule violator, advising him to resolve the problem. In some tribes, a group of fishermen might delegate themselves to speak to the person.
>
> (cited in Singleton 1998: 134)

Among self-organizing forest governance systems, creating and sup-porting a position as guard is frequently essential since resource units are highly valuable and a few hours of stealth can generate sub-stantial illicit income. Monitoring rule conformance among forest users by officially designated and paid guards may make the differ-ence between a resource in good condition and one that has become degraded. In a study of 279 forest *panchayats* in the Kumaon region of India, Agrawal and Yadama (1997) found that the number of months a guard was hired was the most important variable affecting forest conditions. The other variables that affected forest conditions included the number of meetings held by the forest council (a time when infractions are discussed) and the number of residents in the village.

> It is evident from the analysis that the capacity of a forest coun-cil to monitor and impose sanctions on rule-breakers is paramount to maintaining the forest in good condition. Nor should the pres-ence of a guard be taken simply as a formal mechanism that ensures greater protection. It is also an indication of the informal commit-ment of the *panchayat* and the village community to protect their forests. Hiring a guard costs money. The funds have to be gener-ated within the village and earmarked for protection of the resource.

If there was scant interest in protecting the forest, villagers would have little interest in setting aside the money necessary to hire a guard.

(Agrawal and Yadama 1997: 455)

Whether irrigation systems create a formal position as guard depends both on the type of governance of the system and on its size. Of the 15 government-owned irrigation systems included in Tang (1992), 12 or 80 percent have established a position of guard. Stealing water was a problem on most government-owned systems, but it was endemic on the 3 systems without guard. Of the 28 farmer-organized systems, 17 (61 percent) utilize the position of water distributor or guard. Of the 11 farmer-organized systems that do not employ a guard, farmers are vigilant enough in monitoring each other's activities on 5 systems (45 percent), which means that rule conformance is high. That means, of course, that self-monitoring is not high enough on the other 6 systems to support routine conformance with their own rules. A study by Romana de los Reyes (1980) of 51 communal irrigation systems in the Philippines illustrates the effect of size. Of the 30 systems that were less than 50 hectares, only 6 (20 percent) had established a position as guard; of the 11 systems that serve between 50 to 100 hectares, 5 (45 percent) had established guard; and of the 10 systems over 100 hectares, 7 (70 percent) had created guards. She also found that in a survey of over 600 farmers served by these communal irrigation systems, most farmers also patrolled their own canals even when they were patrolled by guards accountable to the farmers for distributing water. Further, the proportion of farmers who report patrolling the canals serving their farms increased to 80 percent on the largest self-organized systems compared to 60 percent on the smallest systems (for an analysis of both farmer-organized and government systems in contemporary Philippines, see Araral 2005).

Boundary and choice rules also affect how easy or difficult it is to monitor activities and impose sanctions on rule infractions. Closing a forest or an inshore fishery for a substantial amount of time, for example, has multiple impacts. It protects particular plants or fish during critical growing periods and allows the entire system time to regenerate without disturbance. Further, during the closed season, rule infractions are highly obvious to anyone as any appropriator in the resource is almost certainly breaking the rules. Similarly, requiring appropriators to use a particular technology may reduce the pressure on the resource, help to solve conflicts among users of incompatible technologies, and also make it very easy to ascertain if rules are being followed. Many irrigation systems set up rotation systems so that only two persons need to monitor actions at any one time and thus keep monitoring costs lower than they would otherwise be. Changing payoff rules is the most direct way of coping with commons dilemmas. In many instances, dilemma games can be transformed into assurance games—a much easier situation to solve.

5.4 Using information, scope, and aggregation rules

These rules tend to be used in ways that complement changes in boundary, choice, payoff and position rules. Individual systems vary radically in regard to the mandatory information that they require. Many smaller and informal systems rely entirely on a voluntary exchange of information and on mutual monitoring. Where resource units are very valuable and the size of the group is larger, more and more requirements are added regarding the information that must be kept by appropriators or their officials. Scope rules are used to limit harvesting activities in some regions that are being treated as refugia. By not allowing any appropriation from these locations, the regenerative capacity of a system can be enhanced. Aggregation rules are used extensively in collective-choice processes and less extensively in operational settings, but one aggregation rule that is found in diverse systems is a requirement that harvesting activities be done in teams. This increases the opportunity for mutual monitoring and reduces the need to hire special guards.

It is important to note that we have not yet found any *particular* rules to have a statistically positive relationship to performance. The essential finding, however, is that the *absence* of any boundary or any choice rules *is* consistently associated with poor performance. Relying on only a single type of rule for an entire set of common-pool resources is also negatively related. As reported above, self-organized irrigation systems do tend on average to have performance levels higher than government-organized systems controlling for physical terrain, but this increased performance level is not due to any specific rules or set of rules that we have yet been able to identify.

6 Viewing policies as experiments

The search for rules that improve the outcomes obtained in commons dilemmas is an incredibly complex task whether undertaken by users or by government officials. It involves a potentially infinite combination of specific rules that could be adopted in any effort to match the rules to the attributes of the resource system itself. To ascertain whether one has found an optimal set of rules to improve the outcomes achieved in a single situation, one would need to analyze how diverse rules affect each of the seven components of such an action situation (or a game) and as a result, the likely effect of a reformed structure on incentives, strategies, and outcomes. Since there are multiple rules that affect each of the seven components, conducting such an analysis would be an incredibly time- and resource-consuming process. For example, if only five changes in rules per component were considered, there would be 5^7 or 75,525 different situations to analyze. This is a gross simplification, however, since some of the important rules used in field settings include more than 25 rules (in the case of boundary rules) and even over 100 variants (in the case of choice rules). Further, how these changes affect the outcomes achieved in a particular location depends on the biophysical characteristics of that location and the type of community relationships that already exist. No set of policy analysts (or even all of the game theorists in the world today) could ever have sufficient time or resources to analyze over 75,000 combinations of rule changes and resulting situations, let alone all of the variance in these situations because of biophysical differences.

Instead of assuming that designing rules that approach optimality, or even improve performance, is a relatively simple analytical task that can be undertaken by distant, objective analysts, we need to understand the policy design process as involving an effort to tinker with a large number of component parts (see Jacob 1977). Those who tinker with any tools, including rules, try to find combinations that work together more effectively than other combinations. Policy changes are experiments based on more or less informed expectations about potential outcomes and the distribution of these outcomes for participants across time and space (Campbell 1969, 1975). Whenever individuals agree to add a rule, change a rule, or adopt someone else's proposed rule set, they are conducting a policy experiment. Further, the complexity of the ever-changing biophysical world combined with the complexity of rule systems means that any proposed rule change faces a nontrivial probability of error.

When there is only a single authority for a large region, policymakers have to experiment simultaneously with *all* of the common-pool resources within their jurisdiction with each policy change. And, once a change has been made and implemented, further changes will not be made rapidly. The process of experimentation will usually be slow, and information about results may be contradictory and difficult to interpret (see Brock and Carpenter 2007). Thus, an experiment that is based on erroneous data about one key structural variable or one false assumption about how actors will react can lead to a very large disaster (see Wilson, Yan, and Wilson 2007). In any design process where there is substantial probability of error, having redundant teams of designers has repeatedly been shown to have considerable advantage (see Landau 1969, 1973; Bendor 1985).

For example, let us imagine a series of inshore fisheries located along the coast of a region and posit that every policy change has a probability of failure of 1/10. If the region were regulated by a single governing agency, one out of ten policy changes would be failures for the entire region. If designing rules were delegated to three genuinely independent authorities, on the other hand, each of these authorities would still face a failure rate of one out of ten. The probability that a failure would simultaneously occur along the entire coast, however, would be reduced from 1/10 to $1/10^3$ or 1/1000. On a coast with many more relatively separable inshore fisheries, the likelihood of a coastal-wide failure is reduced still more. Of course, the failure rate for such design tasks can itself not be known, but the positive effect of parallel, redundant design teams each trying to find the best combination of rules does not depend on any particular error rate. The important point is: If the systems are relatively separable, allocating responsibility for experimenting with rules will not avoid failure, but will drastically reduce the probability of immense failures for an entire region.

7   The advantages of polycentric resource governance systems

The last major task to be undertaken in this chapter is to discuss why a series of nested but relatively autonomous, self-organized, resource governance systems may do a better job in policy experimentation than a single central authority. A polycentric system is one where citizens are able to organize not just one but multiple governing authorities at differing scales (see V. Ostrom, Tiebout, and Warren 1961; V. Ostrom 1991, 1997, 2008). Thus, a polycentric system would have some units at a smaller scale corresponding to the size of the basic common-pool resources in the system. Among the advantages of authorizing the users of smaller-scale common-pool resources to adopt policies regulating the use of common-pool resources are:

- Local knowledge. Appropriators who have lived and appropriated from a resource system over a long period of time have developed relatively accurate mental models of how the biophysical system itself operates, since the very success of their appropriation efforts depends on such knowledge. They also know others living in the area and what norms of behavior are considered appropriate in what circumstances.

- Inclusion of trustworthy participants. Appropriators can devise rules that increase the probability that others will be trustworthy and use reciprocity. This lowers the cost of relying entirely on formal sanctions and paying for extensive guarding.
- Reliance on disaggregated knowledge. Feedback about how the resource system responds to changes in actions of appropriators is provided in a disaggregated way. Fishers are aware, for example, if the size and species distribution of their own catch is changing over time and tend to discuss the size of their catch with other fishers. Irrigators learn whether a particular rotation system allows most farmers to grow the crops they most prefer by examining the resulting productivity of specific fields or talking with others about yields at a weekly market.
- Better-adapted rules. Given the above, appropriators are more likely to craft rules that are better adapted to each of the local common-pool resources than any general system of rules.
- Lower enforcement costs. Since local appropriators have to bear the cost of monitoring, they are more likely than central authorities to craft rules that make infractions obvious to other appropriators so that monitoring costs are lower. Further, by creating rules that are seen as legitimate, rule conformance will tend to be higher.
- Redundancy. The probability of failure throughout a large region is greatly reduced by the establishment of parallel systems of rule making, interpretation, and enforcement (see E. Ostrom 2005 for further elaboration of these elements).

There are, of course, limits to all ways of organizing the governance of common-pool resources. Among the limits of a highly decentralized system are:

- Some appropriators will not organize. While the evidence from the field is that many local appropriators do invest considerable time and energy into their own regulatory efforts, other groups of appropriators do not do so. There appear to be many reasons for why some groups do not organize including the presence of low-cost alternative sources of income and thus a reduced dependency on the resource, considerable conflict among appropriators along multiple dimensions, lack of leadership, and fear of having their efforts overturned by outside authorities.

- Some self-organized efforts will fail. Given the complexity of the task involved in designing rules, some groups will select combinations of rules that generate failure instead of success. They may be unable to adapt rapidly enough to avoid the collapse of a resource system.
- Local tyrannies. Not all self-organized resource governance systems will be organized democratically or rely on the input of most appropriators. Some will be dominated by a local leader or a power elite who only change rules that they think will advantage them still further. This problem is accentuated in locations where the cost of exit is particularly high and reduced where appropriators can leave when local decision makers are not responsible to a wide set of interests.
- Stagnation. Where local ecological systems are characterized by considerable variance, experimentation can produce severe and unexpected results leading appropriators to cling to systems that have worked relatively well in the past and stop innovating long before they have developed rules likely to lead to better outcomes.
- Inappropriate discrimination. The use of identity tags is frequently an essential method for increasing the level of trust and rule conformance. Tags based on ascribed characteristics can, however, be the basis of excluding some individuals from access to sources of productive endeavor that has nothing to do with their trustworthiness.
- Limited access to scientific information. While time and place information may be extensively developed and used, local groups may not have access to scientific knowledge concerning the type of resource system involved.
- Conflict among appropriators. Without access to an external set of conflict-resolution mechanisms, conflict within and across common-pool resource systems can escalate and provoke physical violence. Two or more groups may claim the same territory and may continue to make raids on one another over a very long period of time.
- Inability to cope with larger-scale common-pool resources. Without access to some larger-scale jurisdiction, local appropriators may have substantial difficulties regulating only a part of a larger-scale common-pool resource. They may not be able to exclude others who refused to abide by the rules that a local group would prefer to use. Given this, local appropriators have no incentives to restrict their own use and watch others take away all of the valued resource units that they have not appropriated.

Many of the capabilities of a parallel adaptive system can be retained in a polycentric governance system. Each unit may exercise considerable independence to make and enforce rules within a circumscribed scope of authority for a specified geographical area. In a polycentric system, some units are general-purpose governments while others may be highly specialized. Self-organized resource governance systems, in such a system, may be special districts, private associations, or parts of a local government. These are nested in several levels of general-purpose governments that also provide civil, equity, as well as criminal courts.

In a polycentric system, the users of each common-pool resource would have some authority to make at least some of the rules related to how that particular resource will be utilized, and thus would achieve most of the advantages of utilizing local knowledge, and the redundancy and rapidity of a trial-and-error learning process. On the other hand, problems associated with local tyrannies and inappropriate discrimination can be addressed in larger, general-purpose governmental units who are responsible for protecting the rights of all citizens and for the oversight of appropriate exercises of authority within smaller units of government. It is also possible to make a more effective blend of scientific information with local knowledge where major universities and research stations are located in larger units but have a responsibility to relate recent scientific findings to multiple smaller units within their region. Because polycentric systems have overlapping units, information about what has worked well in one setting can be transmitted to others who may try it out in their settings. Associations of local, resource governance units can be encouraged to speed up the exchange of information about relevant local conditions and about policy experiments that have proved particularly successful. And, when small systems fail, there are larger systems to call upon, and vice versa.

Polycentric systems are themselves complex, adaptive systems without one central authority always dominating all of the others. Thus, there is no guarantee that such systems will find the optimal combination of rules at diverse levels that are optimal for any particular environment. In fact, one should expect that all governance systems will be operating at less than optimal levels given the immense difficulty of fine-tuning any very complex, multitiered system.

Trying to find better ways of overcoming the potential tragedies of the commons is never easy and never finished. With strong empirical evidence that those dependant on small- to medium-size common-pool resources are *not* forever trapped in situations that will only get worse over time, we need to recognize that governance is frequently an adaptive process involving multiple actors at diverse levels. Such systems look terribly messy and hard to understand. The scholars' love of tidiness needs to be resisted. Instead, we need to develop better theories of complex adaptive systems, particularly those that have proved themselves able to utilize renewable natural resources sustainably over time.

Overcoming a commons dilemma is always a struggle (Dietz, Ostrom, and Stern 2003).

References

Acheson, James M., and Roy Gardner (2005) 'Spatial Strategies and Territoriality in the Maine Lobster Industry', *Rationality and Society*, 17(3), 309–341.

Acheson, James M., James A. Wilson, and Robert S. Steneck (1998) 'Managing Chaotic Fisheries' in Fikret Berkes and Carl Folke (eds) *Linking Social and Ecological Systems: Management Practices and Social Mechanisms for Building Resilience* (Cambridge: Cambridge University Press).

Agrawal, Arun (1994) 'Rules, Rule Making, and Rule Breaking: Examining the Fit between Rule Systems and Resource Use' in Elinor Ostrom, Roy Gardner, and James Walker (eds) *Rules, Games, and Common-Pool Resources* (Ann Arbor: University of Michigan Press).

Agrawal, Arun, and G. N. Yadama (1997) 'How Do Local Institutions Mediate Market and Population Pressures on Resources? Forest *Panchayats* in Kumaon, India', *Development and Change*, 28(3), 435–465.

Anderson, Lee G. (1986) *The Economics of Fisheries Management.* Rev. ed. (Baltimore, MD: Johns Hopkins University Press).

Anderson, Lee G. (1992) 'Consideration of the Potential Use of Individual Transferable Quotas in U.S. Fisheries', *The National ITQ Study Report*, vol. 1, 1–71.

Anderson, C., and L. Putterman (2005) 'Do Non-Strategic Sanctions Obey the Law of Demand? The Demand for Punishment in the Voluntary Contribution Mechanism', *Games and Economic Behavior*, 54, 1–24.

Araral, Eduardo (2005) 'Bureaucratic Incentives, Path Dependence, and Foreign Aid: An Empirical Institutional Analysis of Irrigation in the Philippines', *Policy Sciences*, 38(2–3), 131–157.

Baland, J.-M., and J.-P. Platteau (1996) *Halting Degradation of Natural Resources: Is There a Role for Rural Communities?* (Oxford: Clarendon Press).

Bardhan, P. K. (2000) 'Irrigation and Cooperation: An Empirical Analysis of 48 Irrigation Communities in South India', *Economic Development and Cultural Change*, 48(4), 847–865.

Bendor, Jonathan (1985) *Parallel Systems: Redundancy in Government* (Berkeley: University of California Press).

Bendor, Jonathan (1987) 'In Good Times and Bad: Reciprocity in an Uncertain World', *American Journal of Political Science*, 31(3), 531–558.

Bendor, Jonathan, and Dilip Mookherjee (1990) 'Norms, Third-Party Sanctions, and Cooperation', *Journal of Law, Economics, and Organization*, 6, 33–63.

Berkes, Fikret (ed.) (1989) *Common Property Resources: Ecology and Community-Based Sustainable Development* (London: Belhaven Press).

Berkes, Fikret (2007) 'Community-Based Conservation in a Globalized World', *Proceedings of the National Academy of Sciences*, 104(39), 15188–15193.

Berkes, Fikret, David Feeny, Bonnie J. McCay, and James M. Acheson (1989) 'The Benefits of the Commons', *Nature*, 340(6229), 91–93.

Blomquist, William (1992) *Dividing the Waters: Governing Groundwater in Southern California* (San Francisco, CA: ICS Press).

Brock, William A., and Stephen R. Carpenter (2007) 'Panaceas and Diversification of Environmental Policy', *Proceedings of the National Academy of Sciences*, 104(39), 15206–15211.

Bromley, Daniel W., David Feeny, Margaret McKean, Pauline Peters, Jere Gilles, Ronald Oakerson, C. Ford Runge, and James Thomson (eds) (1992) *Making the Commons Work: Theory, Practice, and Policy* (San Francisco, CA: ICS Press).

Campbell, Donald T. (1969) 'Reforms as Experiments', *American Psychologist*, 24(4), 409–429.

Campbell, Donald T. (1975) 'On the Conflicts between Biological and Social Evolution and between Psychology and Moral Tradition', *American Psychologist*, 30(11), 1103–1126.

Cardenas, Juan-Camilo (2000) 'How Do Groups Solve Local Commons Dilemmas? Lessons from Experimental Economics in the Field', *Environment, Development and Sustainability*, 2, 305–322.

Cardenas, Juan-Camilo, John Stranlund, and Cleve Willis (2000) 'Local Environmental Control and Institutional Crowding-Out', *World Development*, 28(10), 1719–1733.

Carpenter, J., P. Matthews, and O. Ong'ong'a (2004) 'Why Punish? Social Reciprocity and the Enforcement of Prosocial Norms', *Journal of Evolutionary Economics*, 14, 407–429.

Casari, Marco, and Charles R. Plott (2003) 'Decentralized Management of Common Property Resources: Experiments with a Centuries-Old Institution', *Journal of Economic Behavior and Organization*, 51, 217–247.

Copes, Parzival (1986) 'A Critical Review of the Individual Quota as a Device in Fisheries Management', *Land Economics*, 62(3), 278–291.

de los Reyes R. P. 1980. *Managing Communal Gravity Systems: Farmers' Approaches and Implications for Program Planning.* Quezon City, Philippines: Ateneo de Manila University, Institute of Philippine Culture.

Dietz, Thomas, Elinor Ostrom, and Paul Stern (2003) 'The Struggle to Govern the Commons', *Science*, 302(5652), 1907–1912.

Elster, Jon (1989) *The Cement of Society: A Study of Social Order* (Cambridge: Cambridge University Press).

Fehr, Ernst, and Simon Gächter (2000) 'Cooperation and Punishment in Public Goods Experiments', *American Economic Review*, 90(4), 980–994.

Fortmann, Louise, and John W. Bruce (eds) (1988) *Whose Trees? Proprietary Dimensions of Forestry* (Boulder, CO: Westview Press).

Frohlich, Norman, Joe A. Oppenheimer, and Cheryl L. Eavey (1987) 'Choices of Principles of Distributive Justice in Experimental Groups', *American Journal of Political Science*, 31(3), 606–636.

Gautam, A. P., and Ganesh Shivakoti (2005) 'Conditions for Successful Local Collective Action in Forestry: Some Evidence from the Hills of Nepal', *Society and Natural Resources*, 18(2), 153–171.

Gibson, Clark, Margaret McKean, and Elinor Ostrom (eds) (2000) *People and Forests: Communities, Institutions, and Governance* (Cambridge, MA: MIT Press).

Gibson, Clark, John Williams, and Elinor Ostrom (2005) 'Local Enforcement and Better Forests', *World Development*, 33(2), 273–284.

Gordon, H. Scott (1954) 'The Economic Theory of a Common Property Resource: The Fishery', *Journal of Political Economy*, 62, 124–142.

Guillet, David W. (1992a) 'Comparative Irrigation Studies: The Órbigo Valley of Spain and the Colca Valley of Perú', *Polígonos*, 2, 141–150.

Guillet, David W. (1992b) *Covering Ground: Communal Water Management and the State in the Peruvian Highlands* (Ann Arbor: University of Michigan Press).

Gupta, Radhika, and Sunandan Tiwari (2002) 'At the Crossroads: Continuity and Change in the Traditional Irrigation Practices of Ladakh'. Paper presented at 'The Commons in an Age of Globalization', the Ninth Conference of the International Association for the Study of Common Property, Victoria Falls, Zimbabwe.

Hardin, Garrett (1968) 'The Tragedy of the Commons', *Science*, 162, 1243–1248.

Hayes, T. M. (2006) 'Parks, People, and Forest Protection: An Institutional Assessment of the Effectiveness of Protected Areas', *World Development*, 34(12), 2064–2075.

Hayes, Tanya, and Elinor Ostrom (2005) 'Conserving the World's Forests: Are Protected Areas the Only Way?', *Indiana Law Review*, 37(3), 595–617.

Jacob, Francois (1977) 'Evolution and Tinkering', *Science*, 196(4295), 1161–1166.

Lam, Wai Fung (1998) *Governing Irrigation Systems in Nepal: Institutions, Infrastructure, and Collective Action* (Oakland, CA: ICS Press).

Landau, Martin (1969) 'Redundancy, Rationality, and the Problem of Duplication and Overlap', *Public Administration Review*, 29(4), 346–358.

Landau, Martin (1973) 'Federalism, Redundancy, and System Reliability', *Publius*, 3(2), 173–196.

Lansing, J. Stephen (1991) *Priests and Programmers: Technologies of Power in the Engineered Landscape of Bali* (Princeton, NJ: Princeton University Press).

Lansing, J. Stephen, and J. N. Kremer (1994) 'Emergent Properties of Balinese Water Temple Networks: Co-Adaption on a Rugged Fitness Landscape' in C. G. Langton (ed.) *Artificial Life III: Studies in the Sciences of Complexity*, vol. XVII (Reading, MA: Addison-Wesley).

Levi, Margaret (1988) *Of Rule and Revenue* (Berkeley: University of California Press).

Maass, Arthur, and Raymond L. Anderson (1986) *... and the Desert Shall Rejoice: Conflict, Growth, and Justice in Arid Environments* (Malabar, FL: R. E. Krieger).

McCay, Bonnie J., and James M. Acheson (1987) *The Question of the Commons: The Culture and Ecology of Communal Resources* (Tucson: University of Arizona Press).

Meinzen-Dick, Ruth (2007) 'Beyond Panaceas in Water Institutions', *Proceedings of the National Academy of Sciences*, 104(39), 15200–15205.

Morrow, Christopher E., and Rebecca Watts Hull (1996) 'Donor-Initiated Common Pool Resource Institutions: The Case of the Yanesha Forestry Cooperative', *World Development*, 24(10), 1641–1657.

Nagendra, Harini, B. Karna, and M. Karmacharya (2005) 'Examining Institutional Change: Social Conflict in Nepal's Leasehold Forestry Programme', *Conservation and Society*, 3(1), 72–91.

Netting, Robert McC. (1974) 'The System Nobody Knows: Village Irrigation in the Swiss Alps' in Theodore E. Downing and McGuire Gibson (eds) *Irrigation's Impact on Society* (Tucson: University of Arizona Press).

Netting, Robert McC. (1993) *Smallholders, Householders: Farm Families and the Ecology of Intensive, Sustainable Agriculture* (Stanford, CA: Stanford University Press).

Ostrom, Elinor (1990) *Governing the Commons: The Evolution of Institutions for Collective Action* (New York: Cambridge University Press).

Ostrom, Elinor (1998) 'A Behavioral Approach to the Rational Choice Theory of Collective Action', *American Political Science Review*, 92(1), 1–22.

Ostrom, Elinor (2001) 'Reformulating the Commons' in Joanna Burger, Elinor Ostrom, Richard B. Norgaard, David Policansky, and Bernard D. Goldstein (eds) *Protecting the Commons: A Framework for Resource Management in the Americas* (Washington, D.C.: Island Press).

Ostrom, Elinor (2005) *Understanding Institutional Diversity* (Princeton, NJ: Princeton University Press).

Ostrom, Elinor, and Harini Nagendra (2006) 'Insights on Linking Forests, Trees, and People from the Air, on the Ground, and in the Laboratory', *Proceedings of the National Academy of Sciences*, 103(51), 19224–19231.

Ostrom, Elinor, and James Walker (1997) 'Neither Markets nor States: Linking Transformation Processes in Collective Action Arenas', in Dennis C. Mueller (ed.) *Perspectives on Public Choice: A Handbook* (Cambridge, MA: Cambridge University Press).

Ostrom, Elinor, James Walker, and Roy Gardner (1992) 'Covenants with and without a Sword: Self-Governance Is Possible', *American Political Science Review*, 86(2), 404–417.

Ostrom, Elinor, Roy Gardner, and James Walker (1994) *Rules, Games, and Common-Pool Resources* (Ann Arbor: University of Michigan Press).

Ostrom, Vincent (1991) *The Meaning of American Federalism: Constituting a Self-Governing Society* (San Francisco, CA: ICS Press).

Ostrom, Vincent (1997) *The Meaning of Democracy and the Vulnerability of Democracies: A Response to Tocqueville's Challenge* (Ann Arbor: University of Michigan Press).

Ostrom, Vincent (2008) *The Political Theory of a Compound Republic: Designing the American Experiment.* 3rd ed. (Lanham, MD: Lexington Books).

Ostrom, Vincent, Charles M. Tiebout, and Robert Warren (1961) 'The Organization of Government in Metropolitan Areas: A Theoretical Inquiry', *American Political Science Review*, 55, 831–842.

Palanisami, K. (1982) *Managing Tank Irrigation Systems: Basic Issues and Implications for Improvement.* Presented at the workshop on Tank Irrigation: Problems and Prospects. Bogor, Indonesia: CIFOR.

Repetto, Robert (1986) *Skimming the Water: Rent-Seeking and the Performance of Public Irrigation Systems.* Research Report no. 4. (Washington, D.C.: World Resources Institute).

Rocco, Elena, and Massimo Warglien (1995) 'Computer Mediated Communication and the Emergence of "Electronic Opportunism".' Venice, Italy: University of Venice, Department of Economics, Laboratory of Experimental Economics.

Schlager, Edella (1990) 'Model Specification and Policy Analysis: The Governance of Coastal Fisheries'. Ph.D. diss., Indiana University.

Schlager, Edella (1994) 'Fishers' Institutional Responses to Common-Pool Resource Dilemmas' in Elinor Ostrom, Roy Gardner, and James Walker (eds) *Rules, Games, and Common-Pool Resources* (Ann Arbor: University of Michigan Press).

Schlager, Edella, William Blomquist, and Shui Yan Tang (1994) 'Mobile Flows, Storage, and Self-Organized Institutions for Governing Common-Pool Resources', *Land Economics*, 70(3), 294–317.

Scott, Anthony D. (1955) 'The Fishery: The Objectives of Sole Ownership', *Journal of Political Economy*, 63, 116–124.

Sherman, K., and T. Laughlin (eds) (1992) *NOAA Technical Memorandum NMFS-F/NEC 91. The Large Marine Ecosystem (LME) Concept and Its Application to Regional Marine Resource Management* (Woods Hole, MA: Northeast Fisheries Science Center).

Shivakoti, Ganesh, and Elinor Ostrom (eds) (2002) *Improving Irrigation Governance and Management in Nepal* (Oakland, CA: ICS Press).

Singleton, Sara (1998) *Constructing Cooperation: The Evolution of Institutions of Comanagement* (Ann Arbor: University of Michigan Press).

Singleton, Sara, and Michael Taylor (1992) 'Common Property Economics: A General Theory and Land Use Applications', *Journal of Theoretical Politics*, 4(3), 309–324.

Spooner, Brian (1974) 'Irrigation and Society: The Iranian Plateau' in Theodore E. Downing and McGuire Gibson (eds) *Irrigation's Impact on Society* (Tucson: University of Arizona Press).

Tang, Shui Yan (1992) *Institutions and Collective Action: Self-Governance in Irrigation* (San Francisco, CA: ICS Press).

Tang, Shui Yan (1994) 'Institutions and Performance in Irrigation Systems' in Elinor Ostrom, Roy Gardner, and James Walker (eds) *Rules, Games, and Common-Pool Resources* (Ann Arbor: University of Michigan Press).

Taylor, Michael (1982) *Community, Anarchy, and Liberty* (New York: Cambridge University Press).

Trawick, Paul B. (2001) 'Successfully Governing the Commons: Principles of Social Organization in an Andean Irrigation System', *Human Ecology*, 29(1), 1–25.

Weinstein, Martin S. (2000) 'Pieces of the Puzzle: Solutions for Community-Based Fisheries Management from Native Canadians, Japanese Cooperatives, and Common Property Researchers', *Georgetown International Environmental Law Review*, 12(2), 375–412.

Weissing, Franz J., and Elinor Ostrom (1991a) 'Crime and Punishment: Further Reflections on the Counterintuitive Results of Mixed Equilibria Games', *Journal of Theoretical Politics*, 3(3), 343–350.

Weissing, Franz J., and Elinor Ostrom (1991b) 'Irrigation Institutions and the Games Irrigators Play: Rule Enforcement without Guards' in Reinhard Selten (ed.) *Game Equilibrium Models II: Methods, Morals, and Markets* (Berlin: Springer-Verlag).

Wilson, James A., James M. Acheson, Mark Metcalfe, and Peter Kleban (1994) 'Chaos, Complexity, and Community Management of Fisheries', *Marine Policy*, 18, 291–305.

Wilson, James A., John French, Peter Kleban, Susan R. McKay, and Ralph Townsend (1991) 'Chaotic Dynamics in a Multiple Species Fishery: A Model of Community Predation', *Ecological Modelling*, 58, 303–322.
Wilson, James, Liying Yan, and Carl Wilson (2007) 'The Precursors of Governance in the Maine Lobster Fishery', *Proceedings of the National Academy of Sciences*, 104(39), 15212–15217.

7
How Does Trade Affect the Environment?

Brian R. Copeland

1 Introduction

History is full of examples of how globalization has affected environmental outcomes. Human migration has profoundly affected the natural environment. Much early trade was commodity based— trade in fish, agriculture, timber, and other raw materials all caused exporting countries to increase their exploitation of the natural environment beyond the level that would have occurred to satisfy local consumption demand. Nevertheless, it is only during the past 20 years that the interaction between trade and the environment has become a subject of sometimes heated public policy debate. This has been motivated by a variety of forces that have recently converged. A growing concern about the seriousness of environmental problems has prompted environmentalists to look at the role of globalization in contributing to pressure on the environment. An increase in the reliance on rules-based institutions to support and manage international trade and investment has increased the scope for conflicts between what used to be thought of as domestic policy (such as environmental regulations) and international trade and investment policy. And the rapid growth of international trade and investment flows has increased concerns about competitiveness and market access, both of which are sometimes seen to conflict with environmental policy.

This chapter provides a review of what we have learned from the recent literature about the interaction between trade and investment liberalization and the environment. The review will not attempt to be fully comprehensive,[1] but will focus on a few major themes in the literature, and try to convey the key ideas and empirical results.

The literature on the interaction between trade and the environment has focussed on several related questions. A central issue is how globalization affects the level and incidence of environmental outcomes. Much of the emphasis has been on whether globalization tends to shift pollution-intensive industry to countries with relatively weak environmental policy (the pollution haven hypothesis). Since environmental outcomes are highly dependent on policy, a second key question has been how environmental policy responds to globalization. Much of this is driven by concerns that more stringent environmental policy will reduce international competitiveness. The "race to the bottom" hypothesis is that competitive pressure will induce governments to weaken environmental policy to shield domestic firms from international competition. But there are also concerns that governments will manipulate (and sometimes tighten up) some types of environmental policy to restrict market access from imports. Finally, there are issues of linkages between trade agreements and environmental policy, such as whether some harmonization of environmental policies is needed, or whether weak environmental policy should be considered an unfair trade subsidy.

The key empirical issue that lies behind many of these questions is whether differences in environmental policy across countries affect trade and investment flows. The pollution haven hypothesis is based on the concern that weak environmental policy attracts pollution-intensive industry. The concerns about competitiveness are based on the idea that stringent environmental policy reduces productivity and drives polluting firms away. The "race to the bottom" fears arise from concerns about competitiveness, and so ultimately depend on how differences in environmental policy affect trade and investment flows. Market access concerns are based on fears that environmental policy can be manipulated to favor domestic firms at the expense of foreigners. It is therefore not surprising that a large literature studies the relation between environmental policy and trade and investment flows. Although difficulties in obtaining good data on environmental policy have constrained the literature, there has nevertheless been significant progress, particularly during the past 10 years.

The theoretical literature on trade and environment has merged standard techniques from both literatures. A pervasive issue that distinguishes this work from the standard trade literature is the endogeneity of environmental policy. Because market failures arising from externalities are fundamental to the analysis of environmental problems, the responses of institutions, norms, and public policy to changes in pressure on the environment are critical to determining outcomes. The "race to the bottom" and market access issues directly focus on endogenous policy responses to trade. But the effect of trade and investment on environmental outcomes also fundamentally depends on the policy regime. Trade-induced environmental damages can be prevented if environmental policy responds to the challenge by tightening up emission standards; but if policy is not responsive, environmental degradation can occur. There has therefore been an emerging literature on the endogeneity of environmental policy. Much of the early stimulus came out of work on the environmental Kuznets curve, where it was noticed that pollution did not necessarily increase with growth, and endogenous policy responses was one of the hypotheses suggested to explain this. There has also been a long-standing effort by researchers from a variety of disciplines aimed at increasing our understanding of how communities manage renewable resources and respond to external changes in markets. And more recently, literature on the political economy of policy has flourished, and some of this work has filtered into the trade and environment literature.

Another important feature of much of the work on trade and the environment has been the increased use of general equilibrium models. Much of environmental economics was developed in a partial equilibrium framework, and while this was suitable when the focus was on local environmental problems, a general equilibrium approach becomes necessary if we want to understand the interaction between trade and the environment. The effect of environmental policy on trade flows is very much an issue of the determinants of comparative advantage, and this needs to be addressed with general equilibrium approach. Moreover, if trade leads to an expansion of the clean sector of an economy, general equilibrium constraints imply a contraction of the pollution-intensive sector. Hence a partial equilibrium focus on the effect of trade on a single sector would not give a complete picture. Income effects are important in determining the stringency of environmental policy, and the effect of trade on income is a general equilibrium effect. And finally there can be important general equilibrium responses to policy that can lead to unanticipated results. The response of goods trade, investment flows, and factor markets to a policy change can be important for the understanding the outcome.

The plan of the chapter is as follows. I first give a brief overview of a framework that will be useful in thinking about how globalization affects environmental outcomes. Section 3 reviews work on competitiveness and the pollution haven hypothesis. In section 4, I consider the impact of trade on natural capital. Section 5 considers consumption-generated pollution, and Section 6 reviews the interaction between globalization and environmental policy.

2   Globalization and environmental outcomes: Overview

Trade and investment affect the environment in a variety of ways.[2] If trade promotes growth, then the *scale* of economy activity increases. All else equal, more consumption and production will tend to generate more environmental damage.

The *composition* of economic activity also changes with trade. Exporting industries will expand; import-competing industries will contract. There are also firm-level effects: the recent international trade literature has emphasized that only the most productive firms tend to export and so trade tends to cause some firms to expand and others to contract or exit. Consumption patterns also change in response to relative price changes and availability of new products. If emission intensities vary across industries, firms, and consumption goods, then these composition effects will have a direct influence on environmental outcomes.

Trade also affects the emission intensity of individual consumer goods and production activities. This *technique* effect may be caused directly by trade: imported consumption goods may have different emission intensities than locally produced goods; and imported technology, raw materials, or intermediate inputs affect producer emission intensities. Emission intensities may also change with the scale of production. And trade may induce endogenous environmental policy responses that induce changes in emission intensities of both consumption and production.

Finally, international trade has direct environmental impacts. Transportation activities generate emissions. And unwanted invasive species have been a consequence of increased trade.

I will not discuss the scale effect in much detail in this review. Economic growth is a policy objective of most countries and especially developing countries; and the desirability of that policy objective is beyond the scope of this chapter. Growth does have environmental consequences, but the debate over what the optimal growth rate should be and whether a consumption path is sustainable goes far beyond the issue of openness to trade. There are large literatures on the relation between trade and growth, and the relation between growth and environmental outcomes. Interested readers are directed to those literatures.[3] Instead

I focus on how openness to trade alters the environmental consequences of the growth path of a country—that is, given levels of income, or given rates of economic growth, is there any reason to think that the environmental consequences of growth are more or less severe for an open economy than for one which is relatively closed? I will therefore focus on composition, technique, and direct effects of trade on the environment.

In thinking about how trade affects environmental outcomes, it is useful to distinguish between pollution generated by consumers and that generated by producers. The key difference here is in how domestic and foreign firms are affected by environmental regulations. If pollution is generated by producers, then a possible response of domestic firms to environmental regulations that are more stringent than those prevailing in other countries is to either concede the market to imports, or to move production to a country with weaker environmental policy. In contrast, when pollution is generated by consumption, all firms wanting to sell in the domestic market are affected by domestic environmental regulations applied to consumption goods. One cannot escape these regulations by shifting production to another country. The distinction between consumption and production as a source of pollution will turn out to be important both in determining how trade affects the environment and in influencing the types of policy responses by governments.

One final classification exercise is useful. Some environmental problems result in a deterioration of natural capital and affect economic productivity. Examples include soil erosion, the depletion of fish stocks either from overfishing or from habitat degradation, and serious human health deterioration from pollution. Other environmental problems reduce the quality of life, but have only small effects on economic productivity. Examples might include water pollution which reduces recreational activities or mild forms of air pollution which causes health problems that do not have significant effects on worker productivity. The distinction here is mainly one of degree—one could argue that virtually all environmental problems ultimately have some effect on natural capital. But the literature has tended to treat these two types of environmental problems separately. Modeling environmental problems of the second type typically proceeds by treating pollution as a "public bad" that has a negative impact on utility of consumers but which does not affect the production frontier. This leads to a policy trade-off between higher income and lower environmental quality. On the other hand, models that capture the effect of environmental problems on natural capital have to explicitly deal with production externalities—increased environmental degradation has a negative effect on production possibilities. With these types of environmental problems, increased environmental degradation can lead to both a long-run decline in real income and a decline in environmental quality. In what follows, I will first review work based on models that do not focus on the role of natural capital; and then in section 4, I will review work that explicitly takes into account natural capital constraints.

3 Competitiveness and the pollution haven hypothesis

In its simplest form, the pollution haven hypothesis is that trade and investment liberalization will cause pollution-intensive industry to shift to countries with relatively weak environmental policy. As Copeland and Taylor (2004) emphasize, however, the literature on the pollution haven hypothesis has actually explored two related hypotheses, and authors have not always taken care to distinguish between the two. The bulk of the literature has tended to focus on the issue of *competitiveness*: do differences in environmental policy affect trade and investment flows? The conceptual experiment here is to ask, given the existing trade regime, whether a tightening of environmental policy will make domestic firms in the affected sector less competitive relative to their foreign competitors. Indications that competitiveness is reduced might include a reduction in exports, an increase in imports, a shifting of production and investment in plant and equipment to locations outside the jurisdiction, or a reduced flow of new investment to the affected region. The second hypothesis, which has received somewhat less empirical attention in the literature, is the strong form of the pollution haven hypothesis. The issue here is whether, given the existing differences in environmental policy across countries, increased openness to trade and/or investment will lead to a shifting of pollution-intensive production to countries with weaker environmental policy. This production shift may either occur via changes in trade patterns, or by changes in flows of capital and direct foreign investment. The difference between the two hypotheses lies in what is being held constant. The competitiveness hypothesis holds the trade regime constant and explores the effect of changing environmental policy. The strong form of the pollution haven hypothesis holds the environmental policy regimes in all countries constant and explores the effects of liberalizing trade or investment.[4]

The competitiveness and pollution haven hypotheses are clearly connected, but are nevertheless different. Evidence in support of the competitiveness hypothesis is necessary for the pollution haven hypothesis to be true, but it is not sufficient. Even if differences in environmental policy do affect trade and investment flows, other factors may matter more and result in trade shifting polluting industry to countries with relatively stringent environmental policy.

Before turning to the empirical work, it is instructive to review what theory has to say about each of these hypotheses. The theoretical case for the competitiveness hypothesis is quite strong and can be illustrated with a simple partial equilibrium model. Figure 7.1 illustrates an import-competing industry in a small open economy facing a fixed world price p^w. The supply from domestic producers is S_0, domestic demand is D, and initially imports are M_0. Now suppose environmental policy is tightened at home, but not in foreign countries. This raises domestic production costs, shifts the supply curve to the left to S_1, and increases imports to M_1. Tightening up domestic environmental policy makes the domestic industry less competitive.

This result is quite robust, as the key assumption it relies on is that domestic environmental policy raises costs at the margin. Nevertheless, there is a competing hypothesis—the Porter Hypothesis (see Porter and van de Linde 1995). Porter's argument is that tighter environmental policy may increase competitiveness. There are couple of ways to overturn our simple result and get a Porter-like result. First, if we move to an

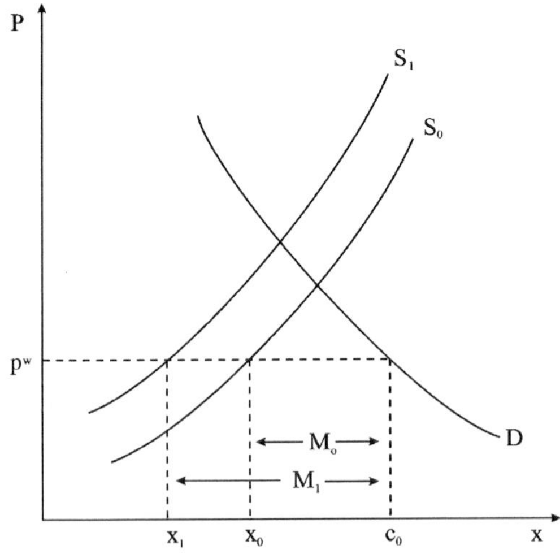

Figure 7.1   More stringent environmental policy reduces competitiveness

imperfectly competitive framework, then it is possible that environmental policy causes a shift to a new technology that has higher fixed costs, but lower marginal costs. This could lead to an increase in a firm's market share (but not necessarily an increase in profits). Second, if there are intermediate goods that affect environmental outcomes (such as an environmental services industry) with either learning or agglomeration effects, then tighter environmental policy can increase the demand for the environmentally friendly intermediate goods and services and, with learning and agglomeration, push down the price of abatement. The net effect on the final good supply reflects the interaction between two effects—complying with environmental regulation is costly, but the average cost of such compliance may fall as everyone else has to comply too and the learning and agglomeration effects kick in (see Greaker 2006). This would tend to mitigate the inward shift of the supply curve.

Let us now turn to the theoretical case for the pollution haven hypothesis. It is a conjecture about how environmental policy affects the pattern of trade, and so we need a more complicated general equilibrium model to address it.

To be concrete, consider a simple generalization of the Heckscher-Ohlin model of international trade, modified to allow for pollution.[5] There are two primary factors of production, capital and labor. X is capital intensive. Let there be two countries, North and South, and suppose North is richer than South.[6] There are two industries: X and Y, and suppose X pollutes during production and Y does not pollute at all. For now, assume that there is no consumption-generated pollution, that pollution reduces utility of consumers but does not reduce productivity in any sector, and that pollution does not spill over international borders. We assume constant returns to scale, identical technologies, identical homothetic preferences over consumption goods, and that preferences are separable with respect to consumption and environmental quality.

Suppose North and South both regulate pollution, and that such regulation is endogenous. For simplicity, we also assume that regulation is perfect; however, the model can be generalized to reflect political economy influences on policy. Since environmental quality is a normal good, one can show that efficient pollution policy becomes more stringent as income rises. Hence we expect North to have more stringent environmental policy than South.

To illustrate the pollution haven hypothesis, first suppose that factor endowment (capital/labor) ratios are the same in North and South. We can use Figure 7.2 to illustrate the pattern of trade. Figure 7.2 illustrates

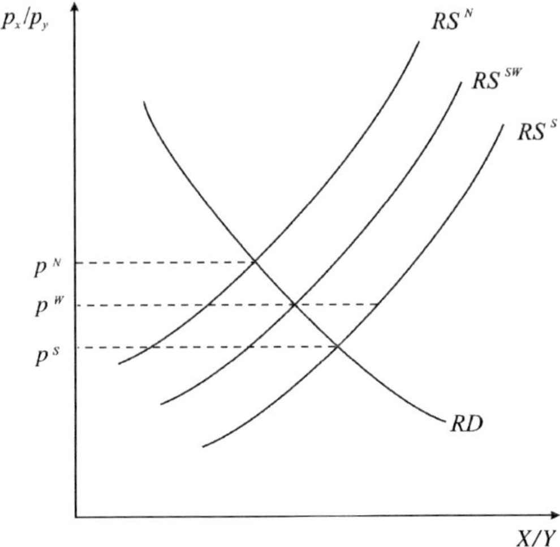

Figure 7.2 Pollution haven with endogenous policy

the demand and supply for X relative to Y. Because we have identical homothetic preferences, the relative demand curve (labelled "RD") is identical in North and South. Moreover, although changes in income will shift the demands for X and Y, they will shift proportionately, so the relative demand curve is not affected by trade-induced changes in income.

If there were no pollution regulation, then because of constant returns to scale and identical technology, North's and South's relative supply curves would be identical (North is just a scaled up version of South in this model). However, because of North's more stringent environmental policy, North's X producers will have to undertake costly measures to reduce emission intensities. This is the competitiveness effect as illustrated in Figure 7.1. Hence for any given $p \equiv p_x/p_y$, North's relative supply curve (RS^N) is to the left of South's (RS^S). In the absence of trade, North's relative price of X is higher than South's ($p^N > p^S$). Because of its weaker environmental policy, South has a comparative advantage in the pollution-intensive good, and it will export the polluting good. To illustrate the free trade equilibrium, I have denoted the world relative supply curve for X/Y by RS^W (it is a weighted average of North's and South's relative supply curves), and so the free trade price is p^W. As trade opens, North reduces its production of the polluting good X and increases its production of the clean good Y. The opposite happens in the South. This model predicts that trade will shift some of the pollution-intensive industry from North to South because of differences in environmental policy.

The effect on environmental quality depends on how environmental policy in each country responds to the change. Copeland and Taylor (2003) show that the changes in environmental policy depend on substitution and income effects and their interaction with the political process. In both countries, the increase in real income arising from freer trade will tend to increase the demand for environmental quality. This, combined with the contraction of X production, leads to a fall in pollution in the North. On the other hand, trading opportunities have increased the marginal benefit of polluting in the South. As Copeland and Taylor show, unless income effects are very strong, trade can be expected to increase pollution in the South, even if environmental policy is efficient. Moreover, because trade shifts polluting industry to the part of the world with the weakest environmental policy, one can show that unless income effects are very strong, trade driven by the pollution haven motive will increase world pollution.

The welfare effects of trade in such a model have been extensively investigated, and depend on the policy regime and whether or not pollution has local or global effects. If environmental policy is suboptimal, there is no presumption that trade liberalization will increase welfare. In the South, the costs of increases in pollution have to be weighed against the income gains from trade. Either gains or losses from trade are possible, depending on the severity of pollution damage costs. However, in Copeland and Taylor (1994), North and South both have perfect environmental policy and both countries gain from trade, despite the increase in pollution in the South. One can think of environmental quality as analogous to a costly input. South is willing to give up some environmental quality in return for the opportunity for higher goods consumption that arises from reallocating its production in response to trade.

If pollution spills over borders and has global effects, the welfare effects can be quite different. Copeland and Taylor (1995) use a model very similar to that outlined above but modified to allow for global pollution. They show that if pollution policy is chosen by each country noncooperatively, then North loses from trade liberalization while South gains. Once pollution is global, South's increase in pollution harms North; and while South is willing to accept higher income in return for diminished environmental quality, North is not.

Although this model focuses on trade, similar logic applies to the case of foreign investment (see Rauscher 1997). North's more stringent environmental regulation acts like a tax on the polluting industry, X. Since X is capital intensive, this means that given any price p, the return to capital is higher in the South than in the North. This leads to a capital outflow from the North to the South, which creates a pollution haven outcome.

Although the model outlined above generates a prediction of a pollution haven, it clearly has some unrealistic aspects. Our pollution haven result was obtained under the assumption that capital/labor ratios were identical across countries. More importantly, the model was set up so that the *only* motive for trade was differences in environmental policy. In reality environmental policy differences are just one of a multitude of motives for trade.

To illustrate how fragile the pollution haven result is, suppose now that North is more capital abundant than South—and let us first think about what would happen if pollution policy were identical across countries. This reduces to the standard Heckscher-Ohlin model of trade. North's capital abundance means that its supply of X relative to Y (RS_0^N) is higher than South's (RS^S) for any given p (recall that X is the capital-intensive industry). This is illustrated in Figure 7.3. The autarky price of the pollution-intensive good is lower in the North than the South. North exports the polluting good and trade shifts pollution from South to North.

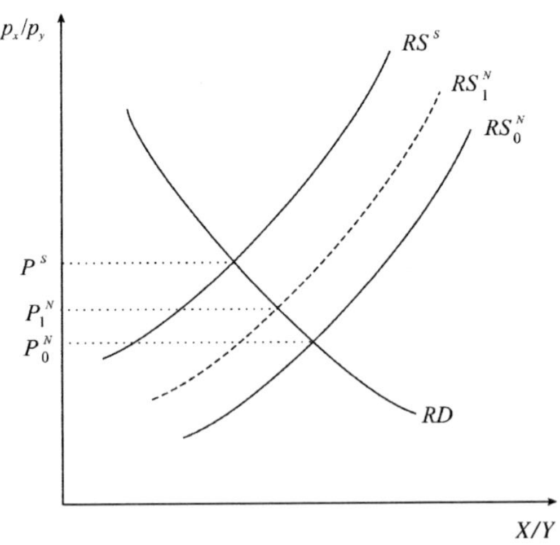

Figure 7.3   Two sources of comparative advantage

Now consider the effect of North's more stringent environmental policy. This shifts North's relative supply curve to the left. But if the effect of capital abundance is stronger than the effect of the differences in pollution policy, then North's relative supply curve would be positioned as illustrated by the dashed line RS_1^N. Despite its more stringent environmental policy, North still has a comparative advantage in the polluting good. Moreover, as Copeland and Taylor (2003) show, in this case, trade can lead to a fall in world pollution by shifting some of the polluting production to the country with relatively more stringent environmental policy.

The point of this example is that in a world where countries differ in many dimensions, relatively weak environmental policy is not enough to attract pollution-intensive industry. Although the discussion above has focused on relative capital abundance, differences in relative supply of other factors of production, infrastructure, institutions, climate, location, and many other factors will affect trade patterns. Market size can also affect trade patterns via agglomeration effects. In the presence of transport costs, many firms find it advantageous to locate near their customers and suppliers. If these agglomeration forces are important, clusters of economic activity will develop which are regions of high income and which place significant pressure on the environment. In response to these pressures, environmental regulations may be quite stringent. But firms may nevertheless continue to locate in such regions to take advantage of the agglomeration benefits. (See Zeng and Zhao 2006 for a model which develops this idea.)

Hence the theoretical support for the pollution haven hypothesis is rather weak. For the pollution haven hypothesis to be true, we need either that the adverse competitiveness effect of stringent environmental policy be strong relative to other motives for trade, or else we need the presence of weak environmental policy to be correlated with other factors that help create a comparative advantage for pollution-intensive industry.

Let us now turn to the empirical work. This work has been reviewed recently by Copeland and Taylor (2004) and by Brunnermeier and Levinson (2004); so I will be concise here. As noted above, much of the empirical literature has focused on the competitiveness hypothesis: does stringent environmental policy affect trade and investment flows? This literature has faced two major challenges.

First, data availability is a serious problem. The key set of data needed is a panel of data on the stringency of environmental policy over time and across jurisdictions. Such data is very hard to come by. Pollution taxes are rarely used and so it is very difficult to obtain clear measures of environmental policy. There are some exceptions. Levinson (1999) studies hazardous waste trade within the United States and uses a series of data on waste disposal charges that vary both across time and across states in the United States. China has used pollution charges quite extensively, and since these charges vary across province, this data has been exploited by some researchers, notably Dean, Lovely, and Wang (2009). Many researchers have used county-level compliance with the US Clean Air Act as a proxy for the stringency of environmental policy at the county level within the United States. The argument here is that counties not in compliance with the Clean Air Act have to have a program of stringent environmental policies to address their non-compliance. Once we move beyond these data sources, though, the quality of data is problematic. Abatement costs have been used by many researchers. However, these have many problems—they are endogenous, they are based on surveys where in many cases it is difficult for respondent to isolate abatement costs from other production costs, and their availability is limited to a very small number of countries. Qualitative measures of the stringency of environmental policy have also been used.

The second challenge faced by the literature on the competitiveness hypothesis is the need to deal with unobserved heterogeneity and endogeneity problems. As an example, consider Figure 7.3 again. A typical test of the competitiveness hypothesis would be to get a series of data on net exports and the hypothesis would be that, all else equal, more stringent environmental policy would tend to reduce net exports. However, note that in the scenario depicted in Figure 7.3, net exports of X (the polluting good) are *positively* correlated with the stringency of environmental policy. North exports the polluting good and South imports it, but North's environmental policy is more stringent that South's. Nevertheless, the competitiveness hypothesis holds in Figure 7.3. Recall that tightening up North's environmental policy caused its relative supply curve to shift in from RS_0^N to RS_1^N. Given the trade regime, this reduced North's net exports of X as predicted by the competitiveness hypothesis. To identify this effect empirically, we need to control for relative capital abundance. However, as discussed above, many factors affect trade patterns, and the researcher is unlikely to be able to control for all of them. Hence we have the problem of unobserved heterogeneity which plagues the early cross-sectional work in this field. During the past 10 years, researchers have used panel data to overcome this problem.

Endogeneity problems arise from a variety of sources. For example, suppose the race to the bottom hypothesis is correct. Then governments would systematically weaken environmental policy in response to competitive pressure from imports. Consequently, we might observe high levels of imports (a lack of competitiveness) correlated with *weak* environmental policy. But in this case it is the trade flows causing the environmental policy, rather than vice versa. Dealing with endogeneity has been problematic. Some researchers have looked for measures of environmental policy that are less likely to be endogenous. The Clean Air Act, for example, is a US Federal policy and so can be treated as exogenous at the county level (although whether or not a county is in compliance may be endogenous). Others have used standard techniques, such as instrumental variables.

Researchers have used data on trade flows, plant location, and foreign investment to test the competitiveness hypothesis. Early work on trade flows used cross-sectional data and tended to find that the stringency of environmental policy had little or no effect on trade flows; and sometimes the sign was opposite to what theory predicted (see Jaffe et al. 1995 for a survey). Researchers in this area built on standard techniques developed in the international trade literature to explain trade flows (such as Leamer 1984 and Baldwin 1971). Tobey (1990) used data on net exports of a set of pollution-intensive goods for 23 countries. He used a set of variables such as factor endowments, tariffs, and an indicator of the stringency of environmental policy to explain trade flows. He found that the environmental policy variables were not significant (however, neither were most of the factor endowment variables). Kalt (1988), Grossman and Krueger (1993), and others regressed US net exports at the sectoral level on industry characteristics (such as labor and capital cost shares, trade policy, share of abatement costs, and other control variables). The abatement cost variable was usually not significant in these studies, although some, such as Kalt (1988), found that higher abatement costs were associated with increased exports.

More recent work using panel data and which takes into account the endogeneity of environmental policy has tended to find support for the competitiveness hypothesis—all else equal, more stringent environment environmental policy tends to reduce net exports. Levinson (1999) used panel data on hazardous waste trade within the United States. Using either state-level fixed effects or correcting for the endogeneity of pollution policy, he found that a higher tax on the disposal of hazardous waste reduces net inflows of hazardous waster into a state. Ederington, Levinson, and Minier (2004) use panel data on US net imports. They find that industry-level abatement costs have a small positive effect on net imports, but the effect is not significant. However, they point out that this is an estimate of the *average* effect on environmental policy on US trade. They argue that the effect would be expected to differ depending on where imports come from or on how footloose production is. They find support for both hypotheses. When they separate their sample in to OECD and non-OECD countries, they find that US environmental policy has a significant positive effect on US imports from non-OECD countries, but not from OECD countries—their interpretation is that differences in environmental policy are bigger in the former case than in the latter. They also construct several measures of how footloose an industry is (transport costs, plant size, etc.) and interact these indicators with abatement costs. The hypothesis is that abatement costs will have a much smaller effect on net trade in industries that are less footloose. They find evidence to support this. Some supporting evidence for the footlooseness hypothesis comes from Gray and Shadbegian (2002) who found evidence that firms in the paper and oil industries in the United States tended to shift production to states with weaker environmental regulation during the period 1967–92. The effect was stronger for the paper industry than for the oil industry. They conjecture that this is because paper is more easily transportable.

There is also evidence that endogeneity of pollution policy matters. Ederington and Minier (2003) and Levinson and Taylor (2008) both study the effects of environmental policy on US net imports at the industry level. Using instrumental variables to correct the endogeneity of environmental policy, they find both statistically and economically significant positive effects of more stringent environmental policy on net imports.

Turning to the work on plant location, early cross-sectional work failed to find support for the competitiveness hypothesis (see Levinson 1996), but recent work using panel data has found that more stringent environmental policy does affect plant location. Becker and Henderson (2000) used data on whether or not a US county was in compliance with the US Clean Air Act as an indicator of the stringency of environmental policy (counties not in compliance are required to have a plan in place to improve air quality, and this entails more stringent environmental policy). Because the policy is imposed at the Federal level, endogeneity problems are mitigated, and because data is available at the county level, they are able to obtain within as well as across state variation. They find that more stringent environmental policy has a significant and sometimes large deterrent effect on the probability that a new plant will locate within a county. Kahn (1997), Greenstone (2002), List et al. (2003), and others have also found similar results.

Work on foreign direct investment has had mixed results. Keller and Levinson (2002) find that abatement costs have a significant negative effect on inward foreign direct investment into US states. However, List, McHone, and Millimet (2004), using data on compliance with the Clean Air Act from New York state, found that while the stringency of environmental regulation did affect plant location decisions of domestic firms, there was no significant effect for foreign firms. Hanna (2006) found that environmental stringency in the United States increases outward investment. She studied the response of US-based multinational firms to the Clean Air Act using a panel of firm-level data from 1966 to 1999. She found that the Clean Air Act caused such firms to increase their foreign assets by 5 percent and their foreign output by 9 percent. In another study of outgoing foreign investment (using sectoral data), Cole and Elliott (2005) found that US abatement costs affect US foreign investment to Brazil and Mexico. On the other hand, Eskeland and Harrison (2003) studied a sample of four developing countries and found no evidence that US abatement costs affected US foreign investment to those countries.

Although much work in this area has focused on the United States because of data availability, the use of pollution charges in China has allowed some researchers to study the effects of differences across Chinese provinces in the stringency of environmental policy on investment flows. Dean, Lovely, and Wang (2009) use data on 2886 manufacturing equity joint ventures during 1993–96 and use charges for water pollution as a province-level measure of the stringency of environmental policy. They find that more stringent environmental policy deters investment from Hong Kong, Taiwan, and Macao in highly polluting industries. However, pollution regulations have no significant effect on location choice for investment from OECD source countries.

Overall, there is some evidence in support of the competitiveness hypothesis. It is strongest for studies based on plant location within the United States, where several studies have agreed that the Clean Air Act has affected plant location, but there is also evidence that environmental policy affects trade flows, especially when endogeneity and heterogeneity across industries are accounted for. The evidence on foreign investment is more mixed, but still quite limited.

It is important to keep in mind that evidence in favor of the competitiveness hypothesis (i.e., evidence that more stringent environmental policy reduces net exports in pollution-intensive industries) should not be interpreted as evidence that environmental policy is welfare-reducing. In the absence of efficient environmental policy, countries will allocate too much of their production to environmentally intensive activities. A move toward efficient environmental policy will reduce pollution-intensive production, and therefore would also be expected to reduce net exports of pollution-intensive goods as the economy shifts toward a cleaner production mix.

There are very few tests of the stronger version of the pollution haven hypothesis. That is, very few studies have attempted to clearly test whether or not trade liberalization or increased openness to investment flows has caused polluting industry to shift to countries with weaker environmental standards. Ederington, Levinson, and Minier (2004) calculate the pollution generated during the production of US manufacturing exports and imports (holding emission intensities constant at 1987 levels for the period 1972–94). They find that US imports have become cleaner (less pollution generated during production) than US exports. This holds for trade with both OECD and non-OECD countries. This is opposite to what is predicted by the pollution haven hypothesis. They also test to see whether the response of imports to tariff reductions is bigger in high abatement cost industries, and find the opposite, which they interpret as evidence that the United States has a comparative advantage in pollution-intensive industries. In terms of the theory sketched above, other factors matter more to trade than pollution regulations. Antweiler et al. (2001) estimate the composition effect of increased openness to trade on SO_2 pollution. They estimate the elasticity of SO_2 pollution with respect to an increase in openness, controlling for the scale of production, proxies for policy, and other factors. They find that this elasticity tends to increase with the per capita income of a country—the pure effect of trade increases SO_2 pollution more in rich countries than poor countries (in fact the elasticities are negative for some poorer countries). Again, this is opposite to what the pollution haven hypothesis predicts. Antweiler et al. suggest that since SO_2-intensive industries are also capital intensive, the effect of capital abundance is much more important than pollution regulation in explaining the effects of trade on SO_2-intensive production. Overall, while there is some evidence supporting the competitiveness hypothesis, there is very little evidence supporting the strong form of the pollution haven hypothesis.

One of the motivating questions that stimulated the pollution haven literature was the question of whether trade is good or bad for the environment. A few papers have attempted to address this question directly, albeit in narrow contexts. Grossman and Krueger (1993) in their original Kuznets curve paper added an openness variable to their equation explaining cross-city levels of air quality. The estimated sign was positive, suggesting that on average, more open economies had better air quality. Antweiler et al. (2001) argued that the effects of openness on environmental quality should vary with country (and city) characteristics since theory predicts polluting industries will expand in some countries and contract in others. They therefore interacted an openness variable with other indicators of comparative advantage in trying to explain SO_2 pollution, and indeed found that the effects vary across countries. More importantly, they explicitly estimated scale, composition, and technique effects of trade. As noted above, they found that the composition effect of trade tended to shift SO_2-intensive industry to rich countries. But these composition effects were found to be quite small. Overall, they found that for the average country in their sample, more open economies tended to have less SO_2 pollution once one added up the scale, composition, and technique effects. Frankel and Rose (2005) take into account the endogeneity of trade and also find that increased trade is associated with lower levels of air pollution.

While these studies are limited in scope, they are consistent with the evidence that differences in pollution policy are not the major determinant of trade patterns. While this work suggests that openness to trade per se is not associated with reduced air quality, some factors that are influenced by the trade regime do affect pollution. Antweiler et al. (2001) find that capital accumulation, the scale of economic activity, and per capita income are all significant determinants of SO_2 pollution. All else equal, more capital-abundant countries and those with a larger scale of economic activity tend to have higher SO_2 pollution, but increases in per capita income tend to be associated with lower pollution (likely via an endogenous policy response). In short, economic growth does affect environmental outcomes, but the available evidence does not support the view that those economies more open to trade have a more polluted growth path than those less open to trade.

4   Natural capital

Most of the empirical work described above and the theory that lies behind it are based on a model of industrial pollution where environmental damages are harmful but do not significantly affect productivity. Much of the literature therefore emphasizes a trade-off between consumption and environmental quality. This approach has been criticized by many (such as Daly 1993) on the grounds that human consumption possibilities are dependent on a healthy environment, so that the consumption/environmental quality trade-off is an illusion and is not sustainable in the long run.

In fact, there is a long tradition of incorporating natural capital into economic models, going back at least as far as Malthus. In the 1950s, Scott Gordon (1954) and Tony Scott (1955) formally modeled market failures associated with natural capital (fisheries in their case) and showed how institutional failures would lead to resource depletion and real income declines. Natural capital constraints have also been incorporated into models of trade and environment, although the empirical literature has lagged the theoretical literature.

Natural capital constraints are critically important for human health and food production; and they are directly relevant for a large fraction of income and trade, especially in developing countries. To think about how natural capital constraints will affect the interaction between globalization and the environment, it is useful to modify the model that we have used above. Suppose there is a natural capital stock N, which affects the production possibilities. Suppose that production in industry X results in degradation of the stock of natural capital N. Natural capital can regenerate on its own, but increased production from X impedes that regeneration. Moreover, think of natural capital N as a public input. This means there is a production externality—each individual producer impacts N, and since N is a public input, other producers are harmed by the environmental degradation caused by any one firm.

It is useful to distinguish between two possibilities depending upon which sectors are reliant upon N for production. The first possibility, which is relevant to renewable resources, is the case where the externality is internal to industry X. That is, productivity in X depends on N, but productivity in another sector Y is not affected by N. The classic example is the fishery, where we can think of N as the stock of fish. Increased harvesting by one firm lowers the stock of fish, which affects productivity of all firms. The second possibility is that the externality is cross-sectoral. That is, industry X causes environmental degradation to N, but this affects productivity in Y (but not X). Examples include the effects of industrial pollution on fisheries, the effects of deforestation on flooding, and effects of pollution on human health (which affects labor productivity). As is apparent, generalizations of this model are possible. Another case would involve all sectors being affected by N (the human health example might fit better in this category); and as well N may be multidimensional (the forestry example might be better captured with a forest stock variable in addition to the environmental services from land that is vulnerable to flooding). However, for clarity I will focus just on these two different cases.

4.1   Renewable resources

Let us first consider the case of renewable resources, where the externality is internal to the sector. I make a simple modification to our earlier model to include natural capital. This model is based on Chichilnisky (1994) and Brander and Taylor (1997a). Let N denote the stock of the resource (and for concreteness, think of fish), and X denote the harvest rate. The natural capital constraint is given by the biology of the resource stock.

$$\dot{N} = G(N) - X \tag{1}$$

All variables are time dependent (time subscripts are suppressed). G is the natural growth function (typically with an inverse U shape). The resource stock grows naturally and reaches a steady state, but increases in harvesting (X) reduce the resource stock.

The production technology for X is:

$$X = F^X(K_X, L_Y, N) \tag{2}$$

and that for Y is

$$Y = F^Y(K_Y, L_Y) \tag{3}$$

Since the production technology for X depends on the resource stock N, increases in harvesting which reduce N ultimately lead to increased costs for other harvesters. If the resource population is not viable once it falls below some threshold level, then it is possible for excessive harvesting to wipe out the stock. On the other hand, productivity in industry Y is not affected by N.

As is well known, outcomes in such markets depend on the conditions of access to the resource and the institutions that are in place to regulate harvesting. We consider two scenarios and then discuss institutional and regulatory issues in more detail later.

Again suppose there are two countries North and South, and to investigate the effects of conservation policy on trade flows, suppose that industry X is open access in the South (producers do not internalize the externality their harvesting imposes on others), but that the management regime in the North is fully efficient, so that the externality is

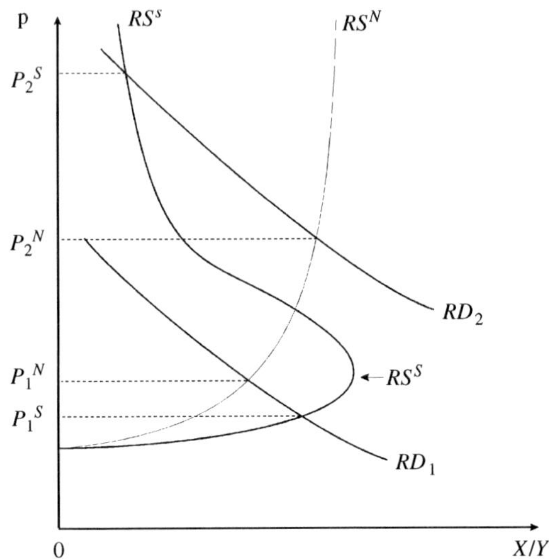

Figure 7.4   Trade and renewable resources

fully internalized (this is the scenario considered in Chichilnisky 1994 and Brander and Taylor 1997b). For simplicity, I also follow Brander and Taylor in assuming that the discount rate approaches zero, so that North's resource managers maximized sustained social surplus.

Figure 7.4 illustrates the relative demand and supply curves. As before, our assumption of identical homothetic preferences means that the relative demand curves are identical across countries. We have illustrated two such relative demand curves here depicting two cases, one of strong demand for X, and the other of weak demand for X.

The relative supply curve for the North (RS^N) looks much the same as a standard relative supply curve. Because the externality is fully internalized, the natural capital constraint essentially acts as an additional constraint on the production function (much like a capacity constraint). Moreover, because we have assumed the discount rate approaches zero, North's managers will ensure that the resource is not depleted in the short run at the cost of lower production in the long run—hence the relative supply curve slopes upward.[7]

The steady state relative supply curve for the South, on the other hand, is backward bending. As the price of X rises, harvesting increases as entrants are drawn into the sector. Hence the relative supply curve initially slopes upward. However, the increase in harvesting depletes the stock (steady state N falls). Eventually, as prices continue to increase, the stock depletion effect dominates and steady state harvest falls. Hence the backward-bending relative supply curve.

Let us now consider the analogue of the pollution haven hypothesis. First consider the case where the demand for the resource is low. This is illustrated by the relative demand curve RD_1 in the figure. In this case, the autarky price of X is higher in North than South. This reflects the tougher conservation measures in the North to preserve the stock. These measures either limit supply or increase harvest costs (such as via harvest taxes or regulations). Consequently, South has a comparative advantage in harvesting. In this case, we get a pollution haven-type result. Trade liberalization will shift some X production from North to South, and increased pressure will be placed on South's environment. Export opportunities amplify the already excessive pressures placed on South's environment.

Although the trade pattern result is similar to that obtained in our earlier pollution haven result, the presence of the natural capital constraint has added one very important difference. In our earlier example, a country that became a pollution haven could lose from trade if pollution policy was too lax. However, in the absence of natural capital, the real income of the economy nevertheless rose. Trade raised real income, but at the cost of a loss in environmental quality. In the current example with natural capital, Chichilnisky (1994) and Brander and Taylor (1997a) show that the South can suffer long-run real income losses from trade liberalization. The short-run export boom induced by trade liberalization leads to a degradation of natural capital which in turn reduces the long-run sustainable level of real income for the economy. Moreover, this real income loss can have multiplier effects. There is much evidence that the stringency of environmental policy is significantly affected by real income levels—increases in income tend to be associated with more stringent environmental policy over time. If trade leads to a loss of long-run real income, this can make it more difficult to maintain and tighten up other environmental regulation, which can further exacerbate environmental problems.[8]

The presence of natural capital has another implication for trade patterns which is significantly different from our earlier pollution example that excluded natural capital. Natural capital is a source of comparative advantage, and since the environmental and conservation policy regime affects the long-run stock of natural capital, the effects of the policy regime on trade patterns are twofold—there is the direct effect of the increased costs (or supply reductions) caused by tighter regulation, but there is also the indirect benefit of a healthier stock of environmental capital if the regulatory regime is successful.

To see this, consider the case of high demand for X. This corresponds to the relative demand curve RD_2 in Figure 7.4. Notice that in this case, the autarky relative price of X is higher in the low-regulation country (South) than in the high-regulation country (North). In this case North's more stringent regulation has given it a comparative advantage (in the long run) in harvesting (this result is due to Brander and Taylor 1997b). The reason for this is that South's weak conservation policy has led to the depletion of natural capital, which has in turn rendered its X industry uncompetitive. It is interesting to note that in this case, despite the absence of effective conservation measures in the South, trade is beneficial for both countries and leads to long-run real income gains for both North and South. Trade takes pressure off the resource in the South and allows its natural capital to begin recovering.

Although empirical work on the effects of trade on natural capital-intensive industries is limited, there are a couple of suggestive examples consistent with this story. Kjaergaard (1994) describes how Denmark ended up in an ecological crisis in the 18th century in large part because of deforestation which led to encroachment by sand dunes, reductions in agricultural productivity, and other problems. International trade played an important role in helping Denmark survive the crisis—imports of wood and alternative fuels took pressure off the domestic forests. More recently, China responded to its problems of deforestation with a logging ban in 1998. A shift to imported wood helped take pressure off Chinese forests.

As is apparent in our North/South example, the institutional and regulatory framework under which resources are harvested is critical for the effects of trade on outcomes. There is significant heterogeneity in resource management institutions and policies across countries and across resources within countries. This suggests that resource management is endogenous and may be responsive to changes in the trade regime. Understanding how resource management is affected by increased trade and investment is critical for understanding outcomes, since, as we discussed above, radically different outcomes can arise depending on the management regime. Trade can produce real income gains without causing resource depletion if a resource is well managed; but it can lead to resource collapse and real income losses if it is poorly managed. Although there has been a great deal of work by Ostrom (1990), Baland and Platteau (1996), and others aimed at trying to explain differences in management approaches, there has been as yet little work that investigates how globalization affects resource management regimes.

Francis (2001), Margolis and Shogren (2002), and Bergeron (2004) develop models in which there is a fixed cost of managing a resource.[9] Once the fixed cost is paid, they assume that management is perfect—externalities are fully internalized. If trade makes the resource more valuable, then it will be worth paying the fixed cost of setting up a management regime. Such models therefore predict that increased export opportunities (which lead to higher prices) can help make the transition to improved resource management. Hotte, Long, and Tian (2000) consider a poaching model; in such models there are variable costs of management. As the resource becomes more valuable, the pressure from poaching increases, but so does the incentive to invest in enforcement to protect the resource.

Copeland and Taylor (2009) endogenize the management regime by developing a model with imperfect monitoring of harvesting. The manager attempts to internalize externalities by restricting harvesting, but individual harvesters have incentives to exceed their quotas. Hence the manager faces incentive constraints. Copeland and Taylor show that the effect of a trade on the effectiveness of the management regime will vary across resources and countries. Some resources will be successful in making the transition to fully efficient management, while other resources will never achieve good management and the higher prices caused by an export boom will lead to resource collapse. The types of economies most vulnerable to trade-induced resource collapse are those with slow-growing resources, impatient agents, very efficient harvesting technologies, and large numbers of agents with a right of access to the resource.

Empirical work on the effects of globalization on the sustainability of renewable resources is still quite sparse. Several studies have found that the enforcement of property rights, corruption, and other measures of institutional quality influence resource depletion (Deacon 1994; Bohn and Deacon 2000; Barbier and Burgess 2001); and Ferreira (2004) found that openness to trade exacerbated deforestation in countries with weak property rights enforcement. There are also case studies that examine particular episodes of trade liberalization in cases where there are weak property rights (see, for example, López 1997, 2000), and as theory would suggest, these find that trade can be good or bad for resource depletion depending on the country's export pattern. There is much scope for further empirical work on this issue.

4.2   Cross-sectoral externalities

In the case of renewable resources discussed above, natural capital can be depleted because of excessive harvesting caused by externalities internal to the sector. However, natural capital can also be depleted because of environmental degradation originating from economic activity outside the sector. Industrial pollution can damage agriculture, forestry, and fisheries; and it can reduce labor productivity throughout the economy via its effects on air and water quality. Deforestation can cause flooding which disrupts many activities. Tourism can be deterred by pollution and wilderness degradation.

To illustrate the implications of cross-sectoral degradation of natural capital, I modify our standard model to incorporate cross-sectoral production externalities. Again suppose there is a stock of natural capital N. It has a natural regeneration rate $G(N)$, but is degraded by industrial pollution Z, which, for simplicity, we assume, is directly proportional to the output of X (hence $Z = \lambda X$)

$$\dot{N} = G(N) - \lambda X \tag{4}$$

The production technology for X is:

$$X = F^X(K_X, L_Y). \tag{5}$$

Industry Y relies on natural capital N and hence we can write its production technology as:

$$Y = F^Y(K_Y, L_Y, N) \tag{6}$$

where F^Y is increasing in N. In this model, increases in industrial output X generate more pollution Z, which reduces the stock of natural capital N, which in turn reduces productivity in Y. Natural capital will be treated as a public input—it is not priced in a market. Hence this model reduces to one with a simple production externality—productivity in sector Y is adversely affected by production in the X sector.

Let us now consider the implications of cross-sectoral externalities for the pattern of trade. Again suppose there are two countries, North and South, with identical homothetic preferences. In this case, we will start by assuming that North and South are completely identical and that neither has any environmental policy. We will then consider the implications of more stringent environmental policy in North.

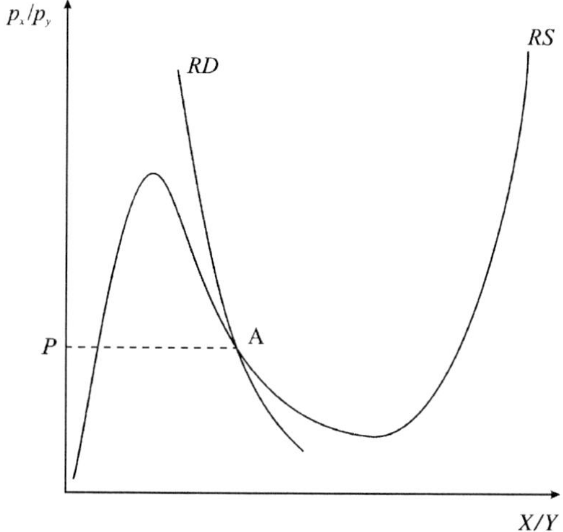

Figure 7.5   Cross-sectoral production externalities

Referring to Figure 7.5, North and South again have identical relative demand curves *RD*. Let us now consider the relative supply curves. I focus here on long-run steady states. The shape of the long-run relative supply curve is quite complex and depends on substitution possibilities between K, L, and N in production. To give a stark indication of the issues, I consider the case explored in Copeland and Taylor (1999) where N is separable from the other inputs in the production of Y, and the capital intensities across X and Y are very similar. In this case, one can show that the long-run (steady state) relative supply curve will be nonmonotonic and will have a region where it is downward sloping, as illustrated in the figure. To see the intuition, think of the relative supply curve as the minimum relative price needed to support a given level of X/Y. For X to increase production relative to Y, it needs to draw resources out of Y, and in the usual model without cross-sectoral externalities, a higher relative price of X is needed to give X producers enough money to lure the factors from Y. However, in the current model, an increase in X production destroys environmental capital and this lowers productivity in Y. Hence an increase in X production can allow X producers to attract factors fleeing a declining Y industry at lower prices than before. Hence increased X production can be supported at a lower price. This result is due to the nonconvexities in the long-run production frontier caused by the cross-sectoral externality (Starrett 1972; Baumol and Bradford 1972).

A downward sloping relative supply curve has a number of interesting implications. First, there is the possibility of multiple equilibria in autarky (this would occur if the relative demand curve was flatter than illustrated in the figure). Countries with very similar endowments and preferences may look quite different prior to trade because of the differences in natural capital. That is, history matters, and can affect comparative advantage via its long-run effects on a country's natural capital.

Second, very similar countries may end up trading and experience very different environmental consequences because of trade. This is the case illustrated in the figure. Point A is a (Marshallian) stable autarky equilibrium.[10] If the countries are identical, then both will have identical prices prior to trade. There is therefore no comparative advantage. However, a no-trade equilibrium is unstable. Suppose trade is free and both countries are at A. Now suppose X producers in the North randomly start producing more X. This will deplete natural capital, which raises costs in Y and creates a comparative advantage for North in X. Hence an initial movement away from point A in North will be reinforced via trade. Moreover, North's movement away from A causes changes in South as well. The flip side of North's comparative advantage in X is a Southern comparative advantage in Y. Hence South starts exporting Y. But this draws resources out of X in the South, which causes its natural capital to regenerate and reinforces its comparative advantage. Hence one can show that the only stable equilibrium in this model involves trade between countries which were initially identical. Moreover, there is a lock-in effect. The changes in natural capital induced by trade tend to amplify differences between countries.

What are the implications for the pollution haven hypothesis? Suppose that North had more stringent environmental policy than South (but that the countries were otherwise identical)—for simplicity think of a quota on X production. This would mean that North had more environmental capital than South prior to trade, and would give North a comparative advantage in Y. This model therefore yields a pollution haven result. Moreover, the presence of natural capital strengthens the pollution haven effects—small changes in policy can lead to lock-in as noted above, so once a country becomes a pollution haven it may be hard to escape. Other differences between countries (such as differences in human capital abundance) can still dominate differences in environmental policy in determining trade patterns. But it is more difficult for these other factors to offset the pollution haven effect once natural capital is depleted.

Finally, this example highlights how the type of natural capital depletion can have a significant effect on how trade affects a country's
environment. Recall that in the case of renewable resource depletion,
trade can actually help a country's environment by taking pressure off
the depleted resource. That is, when the externalities affecting natural capital are internal to an industry, depletion of natural capital can
ultimately create a comparative disadvantage for the industry causing environmental harm. On the other hand, when the externalities
caused by natural capital depletion are cross-sectoral, the depletion of
natural capital actually ends up giving the industry causing environmental harm a comparative advantage. Trade exacerbates the problem
of natural capital depletion in this case.

5   Consumption-generated pollution

Most discussions of the competitiveness and pollution haven hypotheses have tended to focus on production-generated pollution.[11] Much
pollution, however, is generated by consumption. Automobile emissions, home heating, and postconsumer packaging waste are all examples of consumption-generated pollution. Although the line between
consumption and production can be blurred, especially when considering intermediate goods, the key distinction here is that there are some
goods that are traded, but that generate pollution only when used by
final consumers. This means that if consumption is going to occur
within a given jurisdiction, producers cannot avoid the environmental
regulations by shifting production elsewhere. Effective pollution regulations will target any good consumed within the jurisdiction, regardless
of whether it is produced locally or imported.

The effects of pollution regulation on competitiveness can be very
different with consumption-generated pollution than when pollution
is production-generated. In Figure 7.6, I use a simple partial equilibrium model of an import-competing industry to illustrate the issues.
Suppose X generates pollution during consumption. Let D denote the
demand curve.[12] In the absence of pollution regulations, domestic
supply is S_0 and the world price of imports is p_0. Domestic production is initially x_0, consumption is d_0, and imports are initially M_0.
Suppose now that the government imposes a product standard that
reduces the environmental damage caused during consumption. The
same standard applies to all goods consumed in the country, regardless of whether they are imported or produced domestically. Let c
be the (constant) unit cost of complying with the product standard.

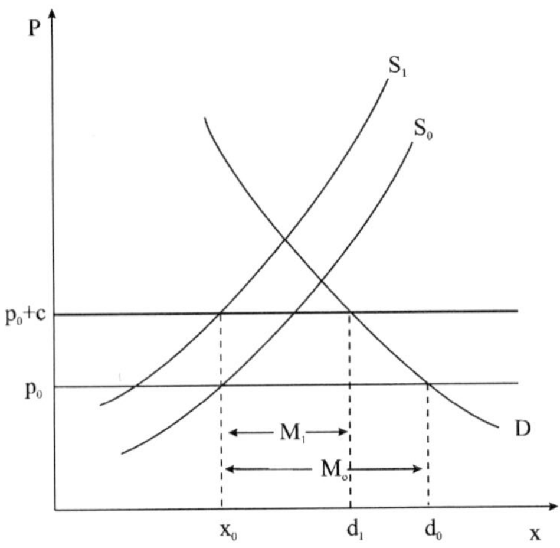

Figure 7.6   Effect of a product standard on competitiveness

Initially assume that this cost is the same for both domestic and foreign producers. Then the world price shifts up to p + c, and the domestic supply curve shifts up vertically by the amount c. Consumption has fallen because the domestic price is now higher. However, notice that the incidence of this decline in sales is borne entirely by foreign producers. Imports fall from M_0 to M_1 but domestic production remains unchanged at x_0. Tougher environmental regulations have not made the domestic industry less competitive than before. Because both foreign and domestic producers have to comply with the environmental regulations, the market price increases to cover the costs of the product standard. There is a "level playing field." In contrast, with production-generated pollution, only the domestic firm has to comply with the regulations, and so the price of imports need not increase in response to new regulations; hence the domestic industry becomes less competitive.

Domestic and foreign compliance costs need not be identical—the example illustrated is only one possibility. Suppose that it is less costly for domestic firms to comply than for foreign firms. Then in Figure 7.6, the product standard would shift up the domestic supply curve by less than the upward shift in the import price. Consequently, tighter environmental regulations would lead to an *increase* in sales by domestic firms. Of course if it were more costly for domestic firms to comply than foreign firms, then domestic sales would fall—however the loss of competitiveness would still not be as significant as in the case of production-generated pollution.

These results have important implications for both empirical work and policy. Work investigating the effects of environmental policy on competitiveness needs to carefully distinguish between consumption- and production-generated pollution, because the predicted effects differ. The policy implications will be discussed in more detail later, but here it suffices to note that firms may sometimes have an incentive to lobby for more stringent environmental regulations because it may give them a cost advantage relative to foreign rivals.

6   How does globalization affect environmental policy?

6.1   Implications for efficient policy

In the classic approach to normative economic policy, openness to international trade and investment does not have a qualitative effect on the design of efficient environmental policy. In a small open economy, the optimal environmental policy in both a closed or open economy requires that environmental externalities be fully internalized (Dixit 1985 remains a classic exposition). In a standard pollution problem, for example, this requires that pollution taxes be set equal to marginal damage. If pollution quotas are instead used, then they should be set so that the (general equilibrium) marginal abatement cost equals marginal damage.

Trade liberalization does not alter the *rule* for setting efficient environmental policy. However, it will alter the *level* of policy. In the case of pollution, efficient policy requires that marginal abatement costs (or marginal benefits of polluting) be equal to marginal damage as illustrated in Figure 7.7. Marginal benefits of polluting depend on goods prices, capital stock, and factor markets. Trade liberalization can cause the marginal benefit curve to shift either in or out. Trade liberalization causing an expansion of pollution-intensive industries either via output price increases, new market opportunities, or foreign investment inflows will shift out the marginal benefit of polluting. On the other hand, if openness causes the polluting sector to contract, or if openness allows access to lower-cost abatement services or technology, then the marginal benefit curve shifts in. The marginal damage curve depends on the characteristics of the pollutant and the harm it causes; and this is influenced by income and prices. If trade increases real income, the marginal damage curve will shift up since environmental quality is a normal good.

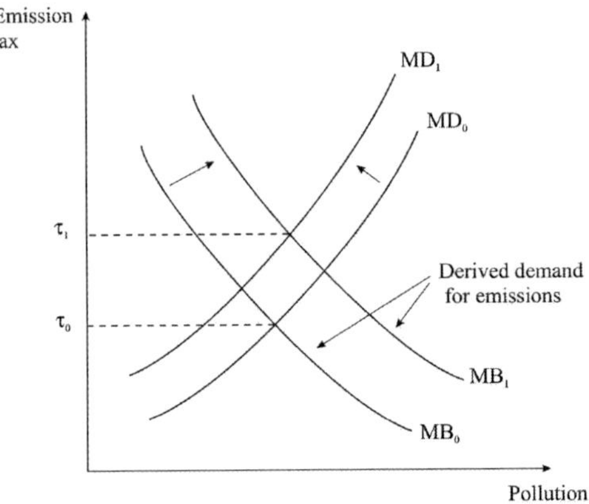

Figure 7.7   Effect of a trade liberalization on pollution tax in export sector

The net effect of trade on pollution levels and the stringency of policy is ambiguous. However, if trade causes an expansion of pollution-intensive activity, then both the marginal benefit and marginal damage curves should shift up and so efficient pollution policy should become more stringent. This implies that it is important that the institutions responsible for setting and enforcing environmental policy be responsive to changes in the trade regime.

If environmental policy is inefficient then the social costs of the inefficiency in policy will be amplified by increased openness. This is most clearly illustrated by thinking about capital mobility. If pollution policy is too weak, then there is excessive production in the pollution-intensive sector. If increased openness attracts foreign investment to that sector, production in the polluting sector will increase and the costs of inefficient policy will be magnified. Formally, this is an application of the Le Chatelier Principle—the response of the polluting sector to changes in the pollution tax from its optimum level is bigger in an economy open to foreign investment (see Copeland 1994; and Neary 2006).

6.2   Race to the bottom: Weak environmental policy as a production subsidy

Much of the concern in policy debates about trade and environment is that governments will not choose to implement efficient environmental policy and that openness to trade and investment will amplify this.

There are several reasons why policy may be inefficient, but here I will focus on two. First, a large country may wish to exert its monopoly and monopsony power in international markets and influence its terms of trade. This can be done by using trade policy; however if free trade agreements preclude the use of trade policy, governments may look for substitutes—manipulating environmental policy is one possibility. Second, environmental policy is influenced by political economy considerations. Governments face pressure to help industries become more competitive, to preserve and promote employment, and to help maintain profits of influential firms. There are many instruments available to do this, but trade agreements tend to reduce the ability of governments to use trade policy and subsidies to achieve these ends. Again, manipulation of environmental policy may be an option.

Both the terms of trade and political economy motives for intervention highlight a key challenge for trade agreements. Trade agreements are incomplete contracts. They restrict explicit barriers to trade, some domestic policy instruments (such as some subsidies), and some overtly discriminatory behavior. But they leave most domestic policy instruments to the discretion of governments. Because any domestic policy that affects either domestic demand or supply has an influence on trade, it can be used as a substitute for trade policy. For example, a government that previously favored an export industry with export subsidies still faces the same incentives to favor the industry after a trade agreement has been signed. All that has changed is the available policy instrument set. Since weak environmental policy is an implicit subsidy, the government may face pressure to weaken environmental policy. This, in essence, is what lies behind the race to the bottom argument.

How well does this argument hold up? The evidence reviewed earlier that environmental policy does affect trade and investment flows suggests that manipulation of environmental is indeed an option for a government that faces pressure to favor a polluting industry. The theoretical case for such intervention is, however, less clear. The results are mixed, and depend on market structure, the type of pollution, and the government's motivation.

First, it is important to distinguish between production- and consumption-generated pollution. This point has been emphasized by McAusland (2008). As was illustrated earlier in Figure 7.6, while more stringent environmental policy reduces competitiveness in the case of production-generated pollution, that need not be the case when there is consumption-generated pollution. Indeed, McAusland shows that a government responding to the interests of polluting firms will weaken environmental policy in the case of production-generated pollution, but may tighten up policy if pollution is consumption-generated. The intuition is based on the asymmetry in the incidence of regulation across the two types of pollutants. Since both domestic and foreign firms are equally affected by environmental regulations targeting consumption-generated pollution, political resistance by producers will be weaker than in the case of production-generated pollution where regulation affects domestic producers but not foreign producers.

Second, environmental policy is only one of many instruments available for governments to protect domestic producers. Although trade agreements may increase the pressure on governments to look for alternative ways of protecting firms, governments also have to be wary of the costs of these alternative instruments. Weaker environmental policy will increase pollution, which increases social costs. These costs have to be weighed against the political costs of other instruments, such as changes in domestic taxes or subsidies, or other regulations that affect the firm. Most of the literature avoids this issue by simply assuming that environmental policy is the only available instrument. This point has been emphasized by Wilson (1996). A couple of recent papers by Sturm (2006) and Kawahara (2008) have taken the multiple instrument issue seriously by using a Coate and Morris (1995) framework. In these papers, consumers will vote out a government that blatantly subsidizes polluting industry. A government captured by domestic pollution-intensive industry may exploit imperfect information about the marginal damage from pollution by weakening environmental policy (and falsely claiming that marginal damage is low).

A third theme in the literature is robustness of results. If governments are motivated by terms of trade concerns, then the optimal trade policy targeting the polluting sector is an import tax if the polluting industry is import-competing, and an export tax if the polluting industry is a net exporter. Hence a government motivated only by terms of trade motives would have incentives to tighten up polluting policy on exporters and weaken it in the import-competing sector. Similar robustness issues arise when firms are imperfectly competitive. Barrett (1994) applies the Brander/Spencer (1985) strategic trade policy argument to pollution policy. In that model, governments have an incentive to provide an export subsidy for exporting firm with market power engaged in Cournot competition with foreign rivals. As Brander and Spencer show, this gives the domestic firm a strategic advantage relative to its foreign rivals, and can shift profits from foreign firms to the domestic firm.

If export subsidies are banned, then Barrett shows weak environmental policy will accomplish the same goal, and in some cases the benefits of this rent-shifting effect can more than offset the cost of increased pollution. However, Barrett also draws on other results in the strategic trade policy literature to show that this result is not robust to different market structures. Although the model predicts weak environmental policy when there is only one domestic firm, it predicts overly strong environmental policy when there are several domestic firms (this is because the terms of trade argument for an export tax dominate the rent-shifting argument once there are more than three domestic firms). And the model predicts excessively weak environmental policy under Cournot competition, but excessively stringent environmental policy when firms engage in price competition. Hence although both the terms of trade and strategic trade policy literature do predict that governments have an incentive to manipulate environmental policy, there is no consensus on whether it would end up being too strong or too weak.

The predictions of the political economy literature are much more robust, at least in the case of production-generated pollution. Polluting firms have incentives to pressure governments for weaker environmental policy to improve their competitiveness. Whether or not the government delivers depends on domestic institutions, the responsiveness of the government to preferences of the consumers, and on which interest groups have relatively strong influence with the government.

The empirical evidence on the responsiveness of environmental policy to trade and investment is limited. Gawande (1999) has fairly convincing evidence that governments do respond to tariff reductions by implementing substitute policies, but he does not focus on environmental policies specifically. Ederington and Minier (2003) find evidence that the stringency of US environmental policy in the manufacturing sector is negatively correlated with import penetration, a result that is consistent with tariff substitution. But this is an issue on which much more empirical work is needed before we can come to conclusions.

6.3   Market access: Environmental policy as a trade barrier

While the "race to the bottom" literature is driven by concerns that trade liberalization may lead to weak environmental policy, there are also concerns that environmental policy may be manipulated (and in some cases set too stringently) to restrict access by foreigners to domestic markets. While it is clear that protectionist governments have incentives to impose more stringent environmental regulations on foreign firms than on domestic firms, trade agreements for the most part prohibit overt discrimination in two ways.

First, there is a distinction between process standards and product standards. Process standards refer to restrictions on how a good is produced (such as emission intensity of production), but which do not affect its final characteristics. Product standards refer to restrictions on the characteristics of a product (such as the emissions generated during consumption). Trade agreements generally do not allow governments to impose process standards or pollution content taxes on imports. That is, the norm is that environmental policy that targets production-generated pollution will vary across countries, and noncompliance with such standards is not legitimate grounds for restricting market access.[13] Such a norm is consistent with efficiency arguments that imply that for pollution with country-specific effects, the optimal environmental standard should vary across countries (because of differences in incomes, local climate, geography, and pressures on the local environment). However, because weak environmental policy is an implicit subsidy to pollution-intensive production, the regime is inconsistent with WTO rules which allow governments to impose countervailing tariffs in response to export subsidies. Nevertheless, implementation of a "green countervail" regime would almost certainly be unworkable because of the difficulty in establishing the efficient environmental policy in every relevant case.[14]

Second, while governments are free to impose product standards, and make compliance with such standards a condition of market access, trade agreements discipline the use of such standards in various ways. The World Trade Organization (WTO) uses a *national treatment* rule—foreign products are expected to be subject to a standard no more stringent than that applied to local products. More recently, the WTO has required that there be some scientific basis for product standards as well. Other trade agreements go farther—in the European Union, for example, there is some centralization and in many cases harmonization in the regulation of product standards.

The analysis of the interaction between market access and environmental policy remains an active area of research. There are two key themes in this literature. The first is that even in a national treatment regime, governments have considerable scope to manipulate standards to favor local producers (because of heterogeneity between local and foreign producers in compliance costs for different types of standards). See, for example, Fischer and Serra (2000), Copeland (2001), McAusland (2004), and Battigalli and Maggi (2003). The second issue is that nondiscrimination policies may constrain governments and make it more difficult to implement appropriate environmental regulation (see, for example, Gulati and Roy 2008). This continues to be a contentious area in the policy realm, since many trade disputes (such as the shrimp-turtle case, restrictions on genetically modified organisms, and the beef hormone case[15]) revolve around tensions between market access and the flexibility of governments to design their own environmental policies.

7   Conclusion

This review of the effects of globalization on the environment has taken a "big picture" focus, asking whether there are systematic reasons why more open economies will have more environmental problems than less open economies. The evidence to date is that while growth and capital accumulation put significant pressure on the natural environment, there is as yet little convincing evidence that openness to trade and investment per se increases environmental damage, on average. However, it is important to keep in mind that behind these averages lie many individual cases where trade will have significant effects on local communities. Trade can threaten the sustainability of renewable resources when the management regime is weak, and the depletion of such resources can have long-lasting negative effects on communities. And while weak environmental policy alone is not the major determinant of trade patterns, it is a contributing factor, and so industries that are both pollution intensive and intensive in unskilled labor will often end up in regions with relatively weak environmental policy.

Another reason not to be too complacent is that a major lesson from the literature on trade and the environment is that active policy intervention matters for environmental outcomes. This is perhaps at the root of some of the tension that arises at times between those working toward liberalized trade and those working for a more sustainable environment. In the trade liberalization agenda, much of the focus is on convincing governments to dismantle policy regimes that they have created to protect local industry. In contrast, the environmental agenda requires the creation and enforcement of policy regimes: setting up tax, quota, and other regulatory mechanisms to internalize externalities and ensure environmentally friendly outcomes. There is no reason in principle for these two agendas to be in conflict—in both cases, the goal is to ensure everyone faces the true social costs of their activities. But in practice, the agendas are different, and this creates challenges in the policy arena.

Notes

1. For other reviews, see Copeland and Taylor (2004), Brunnermeier and Levinson (2004), and Sturm (2003).
2. See Grossman and Krueger (1993) for an early application of the scale, composition and technique decomposition and Copeland and Taylor (1994, 2003) for a theoretical treatment.
3. See, for example, Arrow et al. (2004) and Brock and Taylor (2006).
4. Copeland and Taylor (2004) distinguish between the pollution haven *effect* and the pollution haven *hypothesis*. In that terminology, a pollution haven effect exists if more stringent environmental policy reduces competitiveness in the sense discussed above. The pollution haven hypothesis is defined in the same way as here—it is satisfied if trade or investment liberalization shifts pollution-intensive production to jurisdictions with weak environmental policy.
5. This model is based on Copeland and Taylor (2003), which, in turn, was derived from Copeland and Taylor (1994). See Pethig (1976) for an early pollution haven model with exogenous policy differences.
6. Formally, we can think of the number of agents being the same across countries, but with North's workers having more effective labor (more human capital) and more physical capital.
7. In more general models, with a positive discount rate, resource managers may rationally choose to deplete the stock.
8. On the possibility of a downward ecological spiral, where natural capital depletion reduces income, which leads to more policy failure, which leads to more natural capital degradation, see Daly (1993); for a formal model see Copeland and Taylor (1997).
9. This work has its roots in Demsetz (1967) and the enclosure models of Cohen and Weitzman (1976) and De Meza and Gould (1992).
10. Marshallian stability is the relevant stability condition here because we are considering a long-run equilibrium and it is adjustments in natural capital that lead to a steady state equilibrium. During an out-of-equilibrium adjustment process, markets clear at every point in time, but the level of natural capital may still be evolving. If we consider a price just below that at A, then supply is greater than demand. Hence X production will fall, which causes natural capital to replenish, which moves us back toward A. See Copeland and Taylor (1999).
11. McAusland (2008) notes that of the hundreds of papers available on trade and environment, fewer than a dozen consider consumption-generated pollution.
12. The marginal social benefit of consumption—which I have not illustrated to avoid cluttering the diagram—is lower than demand because of the consumption-generated pollution externality.
13. The issue is more complex in the case of transborder pollution, or in the case of international agreements targeting the environment (such as to protect endangered species). In these cases, governments have sometimes used compliance with environmental standards as a condition of market access.

14. This applies to local pollutants. If there is international agreement on a target for a global pollutant (such as carbon emissions), then a green countervail approach would be much more feasible.
15. See, for example, Weinstein and Charnovitz (2001) for a discussion of these and other cases.

References

Antweiler, W., B.R. Copeland, and M.S. Taylor (2001), 'Is Free Trade Good for the Environment?' *American Economic Review* 91(4): 877–90.

Arrow, K., P. Dasgupta, G. Daily, P. Ehrlich, L. Goulder, G. Heal, S. Levin, K.-G. Maler, S. Schneider, D. Starrett, and B. Walker (2004), 'Are We Consuming Too Much?' *Journal of Economic Perspectives*, Summer.

Baland, J.M. and J.P. Platteau (1996), *Halting Degradation of Natural Resources: Is there a Role for Rural Communities*, Oxford, U.K.: Clarendon Press.

Baldwin, R.E. (1971), 'Determinants of the Commodity Structure of U.S. Trade,' *American Economic Review* 61(1): 126–46.

Barbier, E.B. and J.C. Burgess (2001), 'The Economics of Tropical Deforestation,' *Journal of Economic Surveys* 15(3): 413–33.

Barrett, S. (1994), 'Strategic Environmental Policy and International Trade,' *Journal of Public Economics* 54: 325–38.

Battigalli, P. and G. Maggi (March 2003), 'International Agreement on Product Standards: An Incomplete Contracting Approach,' NBER Working Paper 9533.

Baumol, W.J. and D.F. Bradford (1972), 'Detrimental Externalities and Non-Convexity of the Production Set,' *Economica*, New Series, 39(154): 160–76.

Becker, R. and V. Henderson (2000), 'Effects of Air Quality Regulations on Polluting Industries,' *Journal of Political Economy* 108(2): 379–421.

Bergeron, N. (2004), 'International Trade and Conservation with Costly Resource Management,' mimeo, University of Maryland.

Bohn, H. and R.T. Deacon (2000), 'Ownership Risk, Investment, and the Use of Natural Resources,' *American Economic Review* 90(3): 526–49.

Brander, J.A. and M.S. Taylor (1997a), 'International Trade and Open Access Renewable Resources: The Small Open Economy Case,' *Canadian Journal of Economics* 30(3): 526–52.

Brander, J.A. and M.S. Taylor (1997b), 'International Trade between Consumer and Conservationist Countries,' *Resource and Energy Economics* 19(4): 267–97.

Brock, W.A. and M.S. Taylor (2006), 'Economic Growth and the Environment: A Review of theory and empirics,' in Philippe Aghion and Steven Durlauf (eds), *Handbook of Economic Growth*. Elsevier.

Brunnermeier, S. and A. Levinson (2004), 'Examining the Evidence on Environmental Regulations and Industry Location,' *Journal of the Environment and Development* 13: 6–41.

Chichilnisky, G. (1994), 'North-South Trade and the Global Environment,' *American Economic Review* 84(4): 851–74.

Coate, S. and S. Morris (1995), 'On the Form of Transfers to Special Interests,' *Journal of Political Economy* 103: 1210–35.

Cohen, J.S. and M.L. Weitzman (1975), 'A Marxian Model of Enclosures,' *Journal of Development Economics* 1: 287–336.

Cole, M.A. and R.J.R. Elliott (2005), 'FDI and the Capital Intensity of "Dirty" Sectors: A Missing Piece of the Pollution Haven Puzzle,' *Review of Development Economics* 9: 530–48.

Copeland, B.R. (1994), 'International Trade and the Environment: Policy Reform in a Polluted Small Open Economy,' *Journal of Environmental Economics and Management* 26: 44–65.

Copeland, B.R. (2001), 'Trade and Environment: Product Standards in a National Treatment Regime,' mimeo, University of British Columbia.

Copeland, B.R. and M.S. Taylor (1994), 'North-South Trade and the Environment,' *Quarterly Journal of Economics* 109: 755–87.

Copeland, B.R. and M.S. Taylor (1995), 'Trade and Transboundary Pollution,' *American Economic Review* 85(4): 716–37.

Copeland, B.R. and M.S. Taylor (1997), 'The Trade-Induced Degradation Hypothesis,' *Resource and Energy Economics* 19: 321–44.

Copeland, B.R. and M.S. Taylor (1999), 'Trade, Spatial Separation, and the Environment,' *Journal of International Economics* 47: 137–68.

Copeland, B.R. and M.S. Taylor (2003), *Trade and the Environment: Theory and Evidence*, Princeton: Princeton University Press.

Copeland, B.R. and M.S. Taylor (2004), 'Trade, Growth and The Environment,' *Journal of Economic Literature* 42(1): 7–71.

Copeland, B.R. and M.S. Taylor (2009), 'Trade, Tragedy and the Commons,' *American Economic Review* 99: 725–49.

Daly, H. (1993), 'The Perils of Free Trade,' *Scientific American* 269: 24–9.

Deacon, R.T. (1994), 'Deforestation and the Rule of Law in a Cross-Section of Countries,' *Land Economics* 70(4): 414–30.

Dean, J.M., M.E. Lovely, and H. Wang (2009), 'Are Foreign Investors Attracted to Weak Environmental Regulations? Evaluating the Evidence from China,' *Journal of Development Economics* 90: 1–13.

de Meza, D. and J.R. Gould (1992), 'The Social Efficiency of Private Decisions to Enforce Property Rights,' *Journal of Political Economy* 100: 561–80.

Demsetz, H. (1967), 'Toward a Theory of Property Rights,' *American Economic Review* 57(2): 347–59.

Dixit, A.K. (1985), 'Tax Policy in Open Economies,' in A.J. Auerbach and M. Feldstein (eds), *Handbook of Public Economics*, Vol. 1. Elsevier.

Ederington, W.J. and J. Minier (2003), 'Is Environmental Policy a Secondary Trade Barrier? An Empirical Analysis,' *Canadian Journal of Economics* 36: 137–54.

Ederington, W.J., A. Levinson, and J. Minier (2004), 'Trade Liberalization and Pollution Havens,' *Advances in Economic Analysis and Policy* 4(2), Article 6.

Eskeland, G. and A.E. Harrison (2003), 'Moving to Greener Pastures? Multinationals and the Pollution Haven Hypothesis,' *Journal of Development Economics* 70: 1–23.

Ferreira, S. (2004), 'Deforestation, Openness and Property Rights,' *Land Economics* 80(2): 174–93.

Fischer, R. and P. Serra (2000), 'Standards and Protection,' *Journal of International Economics* 52: 377–400.

Francis, M. (2001), 'Trade and the Adoption of Environmental Standards,' Working Paper, University of Canberra, July 2001.

Frankel, J.A. (2003), 'The Environment and Globalization,' NBER Working Paper 10090.

Frankel, J.A. and A.K. Rose (2005), 'Is Trade Good or Bad for the Environment? Sorting out the Causality,' *Review of Economics and Statistics* 87(1): 85–91.

Gawande, K. (1999), 'Trade Barriers as Outcomes from Two-Stage Games: Evidence,' *Canadian Journal of Economics* 32(4): 1028–56.

Gordon, H.S. (1954), 'The Economic Theory of a Common Property Resource: The Fishery,' *Journal of Political Economy* 63(2): 124–42.

Gray, W.B. and R.J. Shadbegian (2002), 'When do Firms Shift Production Across States to Avoid Environmental Regulation?' NBER Working Paper 8705.

Greaker, M. (2006), 'Spillovers in the Development of New Pollution Abatement Technology: A New Look at the Porter-Hypothesis,' *Journal of Environmental Economics and Management* 52: 411–20.

Greenstone, M. (2002), 'The Impacts of Environmental Regulations on Industrial Activity: Evidence from the 1970 and the 1977 Clean Air Act Amendments and the Census of Manufactures,' *Journal of Political Economy* 110: 1175–219.

Grossman, G.M. and A.B. Krueger (1993), 'Environmental Impacts of a North American Free Trade Agreement,' in Peter M. Garber (ed.), *The Mexico-U.S. Free Trade Agreement*, Cambridge and London: MIT Press, pp. 13–56.

Gulati, S. and D. Roy (2008), 'National Treatment and the Optimal Regulation of Environmental Externalities,' *Canadian Journal of Economics* 41: 1445–71.

Hanna, R. (2006), 'U.S. Environmental Regulation and FDI: Evidence from a Panel of U.S. Based Multinational Firms,' Development Research Institute Working Paper No. 23, New York University, New York.

Hotte, L., N. Van Long, and H. Tian (2000), 'International Trade with Endogenous Enforcement of Property Rights,' *Journal of Development Economics* 62(1): 25–54

Jaffe, A., S. Peterson, P. Portney, and R. Stavins (1995), 'Environmental Regulation and the Competitiveness of U.S. Manufacturing: What Does the Evidence Tell Us?' *The Journal of Economic Literature* 33: 132–63.

Kahn, M. (1997), 'Particulate Pollution Trends in the United States,' *Regional Science and Urban Economics* 27(1): 87–107.

Kalt, J.P. (1988), 'The Impact of Domestic Environmental Regulatory Policies on U.S. International Competitiveness' in A.M. Spence and Heather A. Hazard (eds), *International competitiveness*, Cambridge, MA: Harper and Row, Ballinger, pp. 221–62.

Kawahara, S. (2008), 'Electoral Competition with Environmental Policy as a Second Best Transfer,' mimeo, Fukushima University.

Keller, W. and A. Levinson (2002), 'Pollution Abatement Costs and Foreign Direct Investment Inflows to US States,' *Review of Economics and Statistics* 84: 691–703.

Kjaergaard, T. (1994), 'The Danish Revolution, 1500–1800: An Ecohistorical Interpretation,' *Studies in Environment & History*, Cambridge: Cambridge University Press.

Leamer, E.E. (1984), *Sources of International Comparative Advantage: Theory and Evidence*, MIT Press.

Levinson, A. (1996), 'Environmental Regulations and Industry Location: International and Domestic Evidence,' in Jagdish N. Bhagwati and Robert E. Hudec (eds), *Fair Trade and Harmonization: Prerequisites for Free Trade*, Cambridge, MA: MIT Press, pp. 429–57.

Levinson, A. (1999), 'State Taxes and Interstate Hazardous Waste Shipments,' *American Economic Review* 89: 666–77.

Levinson A. and M.S. Taylor (2008), 'Trade and the Environment: Unmasking the Pollution Haven Effect,' *International Economic Review* 49: 223–54.

List, J.A., W.W. McHone, and D.L. Millimet (2004), 'Effects of Environmental Regulation on Foreign and Domestic Plant Births: Is There a Home Field Advantage?' *Journal of Urban Economics* 56: 303–26.

List, J.A., W.W. McHone, D.L. Millimet, and P.G. Fredriksson (2003), 'Effects of Environmental Regulations on Manufacturing Plant Births: Evidence from a Propensity Score Matching Estimator,' *Review of Economics and Statistics* 85(4): 944–52.

López, R. (1997), 'Environmental Externalities in Traditional Agriculture and the Impact of Trade Liberalization: The Case of Ghana,' *Journal of Development Economics* 53(1): 17–39.

López, R. (2000), 'Trade Reform and Environmental Externalities in General Equilibrium: Analysis for an Archetype Poor Tropical Country,' *Environment and Development Economics* 5(4): 377–404.

Margolis, M. and J. Shogren (2002), 'Unprotected Resources and Voracious World Markets,' *Discussion Paper, Resources for the Future*, Washington DC.

McAusland, C. (2004), 'Environmental Regulation as Export Promotion: Product Standards for Dirty Intermediate Goods,' *Contributions to Economic Analysis & Policy* 3(2).

McAusland, C. (2008), 'Trade, Politics, and the Environment: Tailpipe vs. Smokestack,' *Journal of Environmental Economics and Management* 55: 52–71.

Neary, J.P. (2006), 'International Trade and the Environment: Theoretical and Policy Linkages,' *Environmental and Resource Economics* 33: 95–118.

Ostrom, E. (1990), 'Governing the Commons: The Evolution of Institutions for Collective Action,' Cambridge: Cambridge University Press.

Pethig, R. (1976), 'Pollution, Welfare, and Environmental Policy in the Theory of Comparative Advantage,' *Journal of Environmental Economics and Management* 2: 160–69.

Porter, M.E. and Claas van de Linde (1995), 'Toward a New Conception of the Environment-Competitiveness Relationship,' *Journal of Economic Perspectives* 9(4): 97–118.

Rauscher, M. (1997), *International Trade, Factor Movements, and the Environment*. Oxford: Clarendon Press.

Scott, A. (1955), 'The Fishery: The Objectives of Sole Ownership,' *Journal of Political Economy* 63(2): 116–24.

Siebert, H. (1977), 'Environmental Quality and the Gains from Trade,' *Kyklos* 30: 657–73.

Starrett, D.A. (1972), 'Fundamental Non Convexities in the Theory of Externalities,' *Journal of Economic Theory* 4: 180–99.

Sturm, D.M. (2006), 'Product Standards, Trade Disputes, and Protectionism,' *Canadian Journal of Economics* 39: 564–581.

Tobey, J.A. (1990), 'The Effects of Domestic Environmental Policies on Patterns of World Trade: An Empirical Test,' *Kyklos* 43(2): 191–209.

Weinstein, M.M. and Charnovitz, S. (2001), 'The Greening of the WTO,' *Foreign Affairs* 80: 147–56.

Wilson, John D. (1996), 'Capital Mobility and Environmental Standards: Is There a Theoretical Basis for a Race to the Bottom,' in J.N. Bhagwati and R.E. Hudec (eds), *Fair Trade and Harmonization: Prerequisites for Free Trade*, Cambridge, MA: MIT Press, pp. 393–427.
Zeng, D.-Z. and L. Zhao (2006), 'Pollution Havens and Industrial Agglomeration,' mimeo, Kobe University.

8
Corporate Environmentalism: Doing Well by Being Green

Geoffrey Heal

Introduction

Corporations are often, and quite justifiably, accused of harming the environment. Many of their production processes and products degrade the environment. Yet a certain number of corporations, probably an increasing number, go considerably beyond what is required of them legally in minimizing their environmental impact. They meet legal limits on environmental impacts and then go beyond these. This has been called "overcompliance,"[1] a descriptive, if not elegant, phrase designating going well beyond what is required by laws and regulations in force. Very visible examples are British Petroleum (BP), Starbucks, Heinz, and the banks that have adopted the Equator Principles. In 1997, before the Kyoto Protocol was signed, John Browne, then the CEO of BP, publicly recognized the reality of climate change and the contribution of fossil fuels, and pledged to reduce BP's emissions of greenhouse gases below 1990 levels by 2005. BP met its targets, and clearly deployed considerable managerial resources in doing so. Interestingly, BP claims to have made money from this overcompliance, to the tune of $630 million, mainly through capturing and selling rather than flaring the gases associated with oil fields.[2] Starbucks operates in a very different business, and has also found overcompliance to be worthwhile. Growing coffee on plantations usually requires cutting tropical forests, while the alternative of shade-grown coffee allows the growers to maintain a good fraction of the original forest cover and associated biodiversity. Yielding less per acre, it is a more costly though more environmentally benign way of producing coffee. Starbucks has promoted the sale of shade-grown coffee, and, in conjunction with the non-governmental organization (NGO) Conservation International, worked with coffee growers to teach then how to produce high quality coffee with low environmental impact. There was clearly no legal obligation on Starbucks to do this.[3] In 1990, Heinz encountered criticism for selling tuna caught in a way that killed dolphins, and chose in response to source its tuna in a more expensive but dolphin-friendly way.[4] Again there was clearly no legal pressure to take this action, and no possibility of such pressure. In 2003, a group of large international banks—Citibank, ABN Amro, Barclays, WestLB, and others—agreed to make project finance loans only on projects that meet quite strict social and environmental standards, standards laid out initially by the World Bank and the International Finance Corporation. They require borrowers to have an independent environmental impact assessment of the proposed project, and to agree to an environmental management plan. Failure to comply with this can be seen as a default on the terms of the loan and can lead to the termination of the loan.[5] There are many similar examples,[6] but these suffice to make the point that overcompliance on environmental and social issues is a real phenomenon consuming significant resources at large corporations.

Why do corporations overcomply, going beyond what is legally required of them? The explanation I shall advance here is that they do this to internalize external effects, something that they find in their long-term interests because it reduces the sources of conflicts between them and society. The key point concerns the alignment of corporate and social interests. When there are external effects, the interests of corporations and of society are not aligned: maximizing profits does not lead to the social good. In contrast, in the ideal world of economic theory, with no market failures, maximizing profits leads the economy to a Pareto efficient outcome, which is assumed (indeed defined) to be good for society. A Chief Executive of General Motors, "Engine" Charley Wilson, once said that "What's good for General Motors, is good for America." In a world without market failure, he would have been correct. In the world we live in, he was not, the principal reason being the differences between the private and social costs of making and using automobiles. But by reducing these differences, a company can bring private and social goals into closer alignment. Nonalignment can lead to conflicts with society, often costly and damaging to the corporation. Conflicts can lead to actions against a company by NGOs, to lawsuits, to regulatory intervention, and to loss of brand image and corporate reputation. On the other hand, a reputation for being environmentally considerate can enhance a company's image in the eyes of consumers and improve its relations with regulators. It was actions against Citigroup by an NGO that led to the Equator Principles: Citi was criticized for allegedly making loans on projects that led to deforestation, and the Rainforest Alliance carried out an aggressive campaign trying to persuade customers to end their banking relationships with Citibank because of this—a clear illustration of a private–social cost differential leading to conflicts with potentially costly consequences to Citigroup. Something similar happened with Heinz: They were criticized by environmentalists and then the general public for supporting fishing methods that harmed dolphins, an external effect once again, and chose to adopt instead fishing methods that are "dolphin friendly" and produce no such externality. Interestingly, both BP and Starbucks acted without outside pressures from environmental groups or the public: both acted to forestall such intervention and boost their public images, and have built on these moves extensively in their subsequent promotion campaigns.

Capital markets and externalities

Reducing external effects is not just a matter of improving a firm's image with consumers, but can also affect its market valuation. There is a growing body of empirical evidence that stock markets dislike companies with negative environmental records. The first study to document this was Hamilton's in 1995:[7] he conducted an event study of the first ever release of the US Environmental Protection Agency's (EPA) Toxics Release Inventory (TRI). This is a detailed listing of the emissions of certain toxic chemicals by manufacturing establishments meeting (fairly minimal) size restrictions, and since 1987 has been compiled and made publicly available annually as part of the EPA's "name and shame" campaign to reduce pollution that is not illegal. Hamilton found that featuring on this list had a significant negative impact on share prices, and that the larger the emissions, the more the impact. Subsequent studies have confirmed this effect and found similar effects in other countries.[8]

A recent report by analysts of Union Bank of Switzerland (UBS), apparently unaware of the studies by Hamilton and others, provides an explanation for this reaction on the part of capital markets. In a recent report on corporate social responsibility (CSR),[9] it comments on the connection between social and environmental behavior and the reduction of liabilities: "If a firm or industry 'externalises' costs, the affected stakeholder is very rarely given the opportunity to agree the transfer of costs, and so the 'price' (perhaps very small in the eyes of the firm but very large in the eyes of other stakeholders) is not negotiated at the time when costs are externalised. The danger to firms is that, if the balance of power between stakeholders changes, the price of the exchange may be renegotiated at a future date, and sometimes, but not always, in a court of law." They are arguing here that externalization of costs will generally produce a potential liability to the externalizing company, implying that reducing external costs is a mechanism for reducing potential liabilities. Developing this point further, UBS goes on to comment that "The US Environmental Protection Agency (EPA) has devised a useful definition of a potential environmental liability, which we have adapted here to cover the broader concept of corporate social liability:

- A corporate social liability is an obligation to make a future expenditure due to past or ongoing manufacturing or other commercial activity, which adversely affects any aspect of the environment, the economy, or society.
- A *potential* corporate social liability is a *potential* obligation to make a future expenditure due to past or ongoing manufacturing or other commercial activity, which adversely affects any aspect of the environment, the economy, or society.
- A 'potential corporate social liability' differs from a 'corporate social liability' because an organization may have an opportunity to prevent the liability from occurring by altering its own practices or adopting new practices in order to avoid or reduce adverse environmental, economic or social impacts."

UBS goes on to argue that corporate balance sheets should carry warnings about potential corporate social liabilities, and that valuation exercises by stock market analysts should take these liabilities into account. In this they are close to a recommendation of the UK government, which in a White Paper "Modernising Company Law" published in July 2001 proposed that each company publish every year an Operating and Financial Review (OFR) analyzing and discussing the main factors and trends affecting the company's performance. These would include any social and environmental factors that might affect the shareholders' evaluation of the company's prospects.

Other studies confirm a relationship between environmental performance and financial valuation. Konar and Cohen look at the relationship between the market-to-book ratio, the ratio of the stock market value of the company to the cost of its tangible assets, and a range of environmental factors, including TRI data and environmentally based lawsuits against a company. After allowing for the effects of a broad range of control variables, they find a negative relationship between poor environmental performance and market-to-book. A rather different class of studies of the connection between social, environmental, and financial performance is represented by that of Dowell Hart and Yeung (DHY).[10] Measuring the market-to-book ratio, they found a positive correlation between this and environmental performance. Their study is restricted to US manufacturing companies that are in the S&P 500 and that operate both in the United States and in middle-income developing countries.[11] For the study the authors divided the firms into three categories according to their environmental policies. In the first category were those operating a uniform worldwide standard above that required in the United States. In the second category were those operating at US environmental standards worldwide even if this involves exceeding legally required standards outside the United States, and in the final group were those adopting standards lower than the US in countries where this is permitted. Clearly the first group has the highest environmental standards, and is setting its own worldwide standards above those of the US, which in areas other than greenhouse gas and vehicle emissions are generally the highest. The second group, operating globally at US standards, has the next highest performance and the third group, which is taking advantage of lax local laws in some countries, has the lowest. It is this measure of environmental performance—membership of one of these three groups—that DHY find to be correlated with the ratio of stock market value to the cost of tangible assets. Firms in the first group have higher market-to-book ratios than those in the second, whose market-to-book ratios are in turn on average higher than those of firms in the third group.

The DHY study was pioneering and has justly been the focus of much attention. However, it is important to note that their measure of environmental performance is self-reported and is not independently audited: companies were asked to state which of the three categories they fell into and this statement was not checked. And of course there is the standard comment that correlation does not imply causation, so that the correlation between market-to-book and environmental performance could arise from one or more other factors that are causing both. This is why the event studies of the relationship between stock price movement and the release of information about environmental performance are significant: they can cut through this ambiguity. In spite of these limitations the DHY paper raises interesting questions and is a step forward in connecting one aspect of environmental performance with capital markets and financial performance. One particularly thought-provoking comment by the authors is that capital market valuations internalize externalities—that is, the capital markets recognize difference between private and social costs and treat the excess of social over private as a liability that the corporation will have to meet at some point.[12] This is completely consistent with the findings from the event studies and with the interpretation of assuming social and environmental obligations suggested here.

Fisman Heal and Nair use a rather different set of data. A number of companies make their livings by selling ratings of corporations by their social and environmental performance. One of these is KLD Research and Analytics of Boston. Using data from KLD, Fisman Heal and Nair (FHN) construct three different measures of social performance: one environmental, one related to the treatment of employees, and one based on relationships with the community in which the company operates. The environmental measure reflects pollution, energy use, waste generated, and a range of other activities with environmental impacts. The employee-oriented measure reflects relations with unions, gender and race diversity in the labor force, employee lawsuits, wage levels, and other measures of the treatment of employees. The community measures are based on various measures of giving to the community, support of low-cost housing, and support of educational and cultural objectives. One interesting fact to emerge from this distinction between the different measures of social performance is that firms that rate highly for one type do not necessarily rate highly for others, and indeed in general do not. As we look across different firms we see little correlation between their three scores. Some firms are rated highly on the environmental measure, others on the community measure, and others provide superior treatment of their employees. Few are good at all, and some are good at none.

FHN focus mainly on the community measure, as prior studies have dealt comprehensively with the environmental dimension. We find a correlation between community-oriented performance and market-to-book ratios, even after allowing for differences between firm sizes and for differences between industries. We also conclude that this is more important financially for companies that advertise heavily, suggesting that social performance matters financially most to companies to whom image and visibility are important. We also infer tentatively that the level of social performance relative to other firms in the same industry is more important than the level on its own. This, like the result on advertising and social performance, suggests that consumers are evaluating firms according to their social performance and choosing those with stronger positions. Our finding here is similar to a finding of King and Lennox[13] that a firm's environmental performance relative to the rest of its industry matters for its financial performance. This result is tentative, but is important, as an understanding of how consumers react is of critical importance to firms considering their social policies. When asking what kinds of firms tend to rate highly for social performance, we again find that advertising expenditure is an important variable: firms that spend more on marketing tend to rate higher. This is consistent with the idea that social performance matters for firms for which image and brand reputation are important variables.

The impact of SRI funds

The rapid growth of Socially Responsible Investment (SRI) Funds is an interesting aspect of recent capital market history. The aim of SRI investors is, in some general sense, to use their power in capital markets to do good. And they hope to do well financially in the process. Whether they do is a controversial matter, and not our concern here.[14] Our concern here is with what their impact has been, and whether they have in fact had a positive influence for the causes that they seek to support with their investment strategies. There are three strands of SRI—screened investment, shareholder activism and community investment—and the answers are rather different for each. Puzzlingly, although there is a plethora of studies of the return to screened SRI funds, there is a paucity of studies of their impact. Researchers have either not been interested in whether they have attained their social and environmental goals, or have not seen how to check this. In fact the latter is likely to be the case: it is not easy to see how to check for the impact of SRI funds. By avoiding the shares of certain companies they are shifting demand away from these, and to the extent that share prices depend on supply and demand this may lead to lower prices. Lower share prices will concern managers, partly because they are themselves shareholders, partly because other shareholders will be disturbed and may press for changes, and partly because lower share prices raise the cost of capital to a company. Lower share prices mean that more shares have to be sold to raise any given amount of capital, so that more of the company has to be sold to reach given capital goals. However, it is not obvious that by avoiding certain companies SRI funds will in fact reduce their stock prices. If stock prices depend on expected future earnings, a widely accepted theory of stock prices in the long run, then the fact that SRI funds avoid a company will not affect its stock price, as expected future earnings will not be affected by the funds' behavior. A drop in a price to below expected future earnings because of selling by an SRI fund will just provide an attractive buying opportunity for others in the market. This is not to deny that information about its CSR performance may affect the market's expectations of a company's future earnings.

The studies discussed above have some bearing on this issue. We noted that a company's market-to-book ratio is correlated with its social and environmental ratings. We discussed various explanations of why this might be, including the effects of positive environmental and social behavior on a company's performance. In fact there is another explanation: if a company's market value is correlated with its social and environmental ratings, this could reflect the fact that SRI funds, guided by the SRI ratings, are demanding its shares and inflating its market value. Rather than social and environmental performance raising valuations, it may be that CSR rating acts as a buy signal for SRI funds and raises valuations. So the results we have already seen are consistent with the idea that SRI funds are lowering the cost of capital to highly rated companies, although they certainly do not prove this. In fact if this were the case it would imply that SRI funds are paying above average for their shares and would probably imply lower returns for them in the long run, which does not seem to be the case. If we accept the suggestions that superior social and environmental performance lead to higher stock market valuations, then none of this greater valuation may be attributable to the actions of SRI funds.

The behavior of companies with respect to social and environmental indices such as the Dow Jones Sustainability Index (http://www.sustainability-indexes.com/) and the *Financial Times'* FTSE4GOOD (http://www.ftse.com/ftse4good/index.jsp) provides interesting, if rather casual, data on this point from a different perspective. Both indices are claiming to rate companies according to their attainments in the social and environmental area, broadly interpreted, and both are widely known and very visible, given the families of which they are part. In my experience, many large corporations have been willing to incur significant costs to ensure that they are well-placed on these indices. Presumably this implies that their senior executives see benefits in a clear public recognition of their stature in the social and environmental fields, and when I have spoken with them they have generally explained this in terms of a better position in capital markets and better access to capital, though none have cited hard evidence to support this idea.

There is an interesting recent study that bears directly on the issue of whether SRI funds have an impact on stock prices, suggesting that they do. This paper, appropriately named "The Price of Sin: The Effects of Social Norms on Markets," studies the prices of the "sin stocks" that almost every SRI fund avoids.[15] These are stocks in companies that produce alcoholic drinks or tobacco products, or are active in gambling. Alcohol, tobacco, and gambling are activities that most SRI funds screen against; so if SRI funds have an impact on share prices, then it is likely to be visible in the prices of these stocks. In particular an interesting hypothesis is that if SRI funds are influential then they will tend to depress the prices of sin stocks, so the prices of such stocks will be less than would be expected on the basis of the company's financial performance.

An alternative hypothesis is that their boycotting these stocks will have no effect: to the extent that SRI funds depress the prices of sin stocks than other funds that do not operate ethical screens will find sin stocks attractive buys and will buy enough to bring the price up to the level that their profitability indicates. In other words, the boycotting of these stocks by SRI funds will create arbitrage opportunities for other funds.

Yet another possibility is that the market sees sin stocks as more risky than the average because of the risk of litigation: this has certainly been a factor for tobacco firms in the last two decades. Perceived riskiness will lower a stock's price. The authors, Hong and Kacperczyk (HK), check all of these ideas carefully. Specifically, HK test the following hypotheses: that fewer institutional investors hold sin stocks than other comparable stocks, that fewer analysts cover such stocks than comparable stocks, that the market values of sin stocks are lower than what should be expected from their financial characteristics, and that companies whose stocks are sin stocks rely more on debt financing than comparable companies. Their data set supports all of these suggestions.

The number of institutional investors holding sin stocks is less than the average, as is the number of analysts who report on such stocks. So they form a relatively neglected part of the market. HK's findings on the pricing of sin stocks are particularly interesting. Sin stocks behave like value stocks—that is, stocks that are underappreciated and undervalued by the stock market. Stocks that are undervalued often perform well as they tend to catch up to the rest of the market, and this is what HK find for sin stocks. Their prices are low but the total return to holding them is above average. This is good for investors but of course bad for the issuers, and as a consequence companies in the sin businesses tend to raise less money on the stock market and more on the debt market than comparable companies: they are in financial terms more highly leveraged with a higher debt to equity ratio. The authors also try to understand why sin stocks offer a higher return than others. One possible explanation is that they are seen as more risky, because of the chance of product liability litigation. Stocks that are more risky than the average have to offer a higher return than the average to find buyers. Another explanation is that sin stocks are undervalued just because they are overlooked: some investors are not interested and relatively few analysts cover them. HK decide in favor of the latter explanation. In this they are guided by the fact that after the tobacco settlements of the late 1990s tobacco companies were not at risk for further litigation, as claims against them were settled, yet this did not change their market behavior.

The HK study is the first to give a clear answer to the question: do SRI funds matter? The answer is a limited yes. We still do not know if the prices of "good" stocks are helped by the activities of SRI funds, though we do know that being green helps a company's stock prices—but not necessarily because of the actions of SRI funds. But we do now know that SRI funds have a far-reaching effect on the issuers of sin stocks, affecting their stock prices, who owns them, who follows them, and the companies' financial structures. So even if they do not help firms that "do good," SRI funds may punish the sinners.

When it comes to shareholder activism, matters are much clearer. Corporate law in the United States, and indeed in most countries, allows shareholders with a minimal stake in a company ($2000 in the United States) to place items on the agenda of a shareholder meeting, to place a 500-word supporting statement in proxy statement distributed before the meeting, and to require that a vote be taken on these matters at meeting (the vote is not binding on the company). This is a powerful mechanism for embarrassing management about alleged ethical failures. The annual meetings of large corporations receive wide press coverage and these critical resolutions produce negative publicity, possibly leading to boycotts and diminished retail sales. Shareholder advocacy has been used by large institutional investors, such as the California Public Employees' Retirement System (CaLPERS) and the College Retirement Equities Fund (CREF) in the United States, as a route to more open corporate governance. Large investors have tried to influence corporate policies on such matters as chief executive succession, board membership, and poison pills, although their success rate is not clear.[16] A small fraction of the resolutions submitted lead to the adoption of the recommended policy by the target corporation, although this statistic could be misleading because in some cases, whose number is not known, the institutional investor will approach the corporation before submitting a resolution to see if an agreement can be reached without public debate. There is evidence that the largest institutions have a higher success rate in these nonconfrontational approaches than they do through formal resolutions, perhaps not surprising given that formal resolutions will often be submitted only after lower-key approaches have been tried and have failed. Two informational intermediaries play an interesting role in the process of voting on shareholder resolutions, Institutional Shareholder Services (ISS) and the Investor Responsibility Research Center (IRRC).[17] These groups research the issues that arise in shareholder resolutions and make recommendations to institutional shareholders on how to vote. Their recommendations have been influential with institutional investors, and both have been paying more attention to issues relating to CSR in recent years.

Ethical investors can and do use this same route. According to a report by the Interfaith Center on Corporate Responsibility, in 1999 SRI managers filed about 220 shareholder resolutions with more than 150 US companies. The largest number covered environmental issues, with equity and corporate responsibility taking the next two places. Most of these resolutions are not passed by the shareholders—and even if they were, they would not be binding on the corporation. But the aim is not to pass resolutions: it is to get an issue on the agenda of the Board of Directors, and to start the company thinking about it. The proponents of the resolution see this as the start of a dialogue that may last years before it is productive, although there have been occasions on which shareholder activists find themselves knocking on an open door. A notable case of this type was the decision by Home Depot, a major US Do-It-Yourself outlet, to stop buying mature wood from endangered forests. In this case, shareholder activism was accompanied by a consumer boycott organized by rainforest-related NGOs. Baxter International, a maker of health care products, also agreed to stop using polyvinyl chloride (PVC) in some of its products. PVC releases carcinogens when it is burnt. Chevron and Exxon are facing similar actions by environmental NGOs intended to force them to abandon plans to drill in the Alaskan Arctic wildlife refuge. Another interesting achievement of shareholder activism can be seen in a project run jointly by two major brand names, Disney and McDonald's. McDonald's has exclusive restaurant industry marketing rights to Disney properties, including film, home video, theme parks, and television, so that the two are in effect running a joint venture in the manufacture of Disney items for sale in the branches of McDonald's. At the instigation of several faith-based investment funds that are shareholders in both groups, and in collaboration with these groups, the two companies are investing considerable effort and resources in monitoring the labor conditions under which these products are made. This is not an easy undertaking: many companies have gone public with the problems they have encountered with ensuring compliance with labor standards in China. In this process they have enlisted the help of Chinese groups that are also concerned about labor standards.

Consumer responses to environmental issues

Consumer responses to a company's environmental and social stances can affect their purchasing choices. A very elegant illustration of this was provided by an experiment organized by Hiscock and Smyth at the ABC Department Store in Manhattan. ABC is a rather upmarket department store in Manhattan, itself an upscale location, so that the generality of this experiment is probably limited. Nevertheless it is thought-provoking. The experimenters found two competing ranges of towels, both made in developing countries of organic cotton and under fair trade conditions. Both were therefore exemplary from social and environmental perspectives, but neither was labeled so in the store. The experimenters first labeled one set of towels to indicate its social and environmental credentials, and noted the effects on sales. They were dramatic: sales of the labeled brand rose over those of their competitors. Higher sales persisted even when the prices of the labeled items were increased by 10 percent, and sales began to fall back to the original levels only when prices were raised as much as 20 percent. Clearly consumers were voting with their dollars for products with a positive social and environmental angle. This conclusion is reinforced by a subsequent rerun of the experiment: after the first round all labels were removed and the towels were left unlabeled as initially. After a few months the experiment was reversed—the previously unlabeled towels were now labeled as organic and fair trade while the others remained in anonymity. Again sales of the labeled towels took off.

So there clearly are consumers who judge products partly by their social and environmental credentials, which can therefore be a aid in marketing these products. The experience of the outdoor clothing brand Patagonia in introducing organic cotton, which necessitated a price rise, confirms this: they found no loss of sales in response to a carefully explained replacement of regular by organic cotton and a simultaneous price rise of about 10 percent. These findings are consistent with the FHN findings mentioned above, which indicate that socially responsible behavior can help the valuation of companies that spend heavily on promotion and for which image presumably matters. Indeed it may be behind some of the findings that environmentally responsible behavior is correlated with high market-to-book ratios.

Unfortunately consumers are rarely well-informed about the environmental characteristics of the products available to them, so this chain of thinking suggests a possible role for better information in this field. Clearly good companies have every incentive to represent themselves as such, but of course so do bad ones, and consumers do not have any obvious way of discriminating. It is possible that some aspects of social or environmental behavior can emerge as signals that discriminate between the genuine and the "green-washers," as suggested by Milgrom and Roberts and FHN, but there is clearly a role for third-party certification systems, which, interestingly, have begun to emerge. The Forest Stewardship Council (FSC) and Marine Stewardship Council (MSC) are third-party independent agencies that certify that wood or fish respectively are sustainably harvested. Until very recently they had little leverage, but within the last year some high-profile corporations have adopted them and will make them more widely known. For example, Wal-Mart recently announced that within 5 years it would sell fish only if certified as sustainably caught by the MSC. As Wal-Mart is the largest fish retailer in the United States, this is a significant step and will give the MSC additional significance. Unilever, one of the largest vendors of fish products, has already committed to using only MSC-certified products.

Conclusions

Corporations often go beyond what is legally required when it comes to protecting the environment. There are many well-documented cases in which they are clearly incurring significant costs to do this. Such behavior requires an explanation. My suggestion is that they find it in their own long-term interests to reduce the potential for conflicts between themselves and the rest of society, and seek to do this *inter alia* by reducing external effects, the classical purveyors of environmental damage. In so doing they may be rewarded by the stock market, which seems averse to companies with bad environmental records. As part of this phenomenon, they may avoid the attentions of socially responsible investors, whose boycotting of stocks seems capable of producing undervaluation. They may also be rewarded by consumers, who are clearly in some cases willing to pay extra for products whose social and environmental credentials are clear to them.

Notes

1. By Forest Reinhardt in "Market Failures and the Environmental Policies of Firms: Economic Rationales for 'Beyond Compliance' Behavior," Journal of Industrial Ecology 3 No 1 (Winter 1999) 9–21.
2. Forest Rainhardt, "Global Climate Change and B.P. Amoco," Harvard Business School Case (April 2, 2000).
3. James E. Austin and Cate Reavis, "Starbuck and Conservation International," Harvard Business School Case (October 2, 2002).
4. Richard H.K. Vietor, Forest Reinhardt and Peggy Duxburym, "Starkist (A)," Harvard Business School Case (April 22, 1994).
5. See Geoffrey Heal, *When Principles Pay: Corporate Social Responsibility and the Bottom Line*, Columbia Business School Press, 2008.
6. See Geoffrey Heal, *When Principles Pay: Corporate Social Responsibility and the Bottom Line*, Columbia Business School Press, 2008.
7. J.T. Hamilton, "Pollution as news: Media and stock market reactions to the toxics release inventory data," Journal of Environmental Economics and Management Volume 28, Issue 1, 1995, pp. 98–113.
8. See Geoffrey Heal, *When Principles Pay: Corporate Social Responsibility and the Bottom Line*, Columbia Business School Press, 2008.
9. See Geoffrey Heal, *When Principles Pay: Corporate Social Responsibility and the Bottom Line*, Columbia Business School Press, 2008.
10. Glenn Dowell, Stuart Hart and Bernard Yeung, "Do Corporate Global Environmental Standards Destroy or Create Market Value?" Management Science 46, No 8 (August 2000) 1059–74.
11. Middle-income developing countries are a category defined by the World Bank and consist of those countries with income levels between $1000 and $11000.
12. Interestingly, this is exactly how financial analysts assessed the appropriateness of the drop in Merck's share price after the withdrawal of Vioxx—they calculated the loss of profits and then also the legal liability to which Merck was exposed because of the costs possibly imposed on the users of its product Vioxx.
13. Andrew A. King and Michael J. Lennox, "Does it Really Pay to be Green? An empirical Study of Firms' Environmental and Social Performance," Journal of Industrial Ecology, 5, No 1 (2001), 105–116.
14. See Geoffrey Heal, *When Principles Pay: Corporate Social Responsibility and the Bottom Line*, Columbia Business School Press, 2008.
15. By Harrison Hong of Princeton University and Martin Kacperczyk of the University of British Columbia. Available at http://www.princeton.edu/~hhong/priceofsin1105.pdf.

16. See 'Shareholder Activism and Corporate Governance in the United States,' Bernard S. Black, Professor of Law. Columbia University, in Peter Newman (ed.) *The New Palgrave Dictionary of Economics and the Law* (1998) and 'A Survey of Shareholder Activism: Motivation and Empirical Evidence' by Stuart L. Gillan and Laura T. Starks in *Contemporary Finance Digest* (Autumn 1998) 2, 3: 10–34.
17. ISS recently purchased IRRC.

Author Index

Subject Index